FROM COTSWOLDS
TO
HIGH SIERRAS

Andre & Andre, Las Vegas, Nevada
GEORGE E. FRANKLIN

PLATE I

FROM COTSWOLDS
TO
HIGH SIERRAS

By

GEORGE E. FRANKLIN

ILLUSTRATED WITH PHOTOGRAPHS

THE CAXTON PRINTERS, LTD.
CALDWELL, IDAHO
1966

Library of Congress Catalog Card No. 66-20373

Printed, lithographed, and bound in the United States of America by
The CAXTON PRINTERS, Ltd.
Caldwell, Idaho 83605
105102

TO
MY WIFE FLORENCE, MY CONSTANT
COMPANION AND INSPIRATION

Acknowledgments

The author wishes to express his gratitude to Virginia Kellog and Florence Lee Jones for constructive criticism, helpful suggestions, and encouragement.

Epic of the American Horse, by Jack C. Christie, and the *Encyclopaedia Britannica* proved most helpful in my research relative to American horses.

Foreword

THIS BOOK IS WRITTEN IN COMPLIANCE WITH THE RE-
quests of several of my contemporaries with whom I
have exchanged experiences and anecdotes. It may not
be chronologically correct, and the names of persons and
places may be changed since it is difficult to remember
over such a long period of time. It is, however, sub-
stantially as it happened.

These experiences and adventures cover a life span
of seventy-four years and range from a small village in
the Cotswold Hills of England to the western United
States. They tell how a boy's life, even in staid old England, may be interesting and adventuresome, and how he
may, if he has the intestinal fortitude, travel at the age
of twelve from England to the state of Idaho and adjust
to the life of the West as it was in 1903.

Through the years I have experienced bitter disappointment as well as the sweet pleasure of achievement,
and, since most of the events related here have occurred
between my boyhood home in the Cotswold Hills of
England and the high mountains of the Western United
States, I shall call this book FROM COTSWOLDS TO HIGH
SIERRAS.

Contents

Illustrations

FROM COTSWOLDS
TO
HIGH SIERRAS

England

I WAS BORN IN A SMALL VILLAGE IN THE COTSWOLD Hills of England on the nineteenth day of June, 1890. On the twelfth of July, 1891, my father died, so I never knew him. I have regretted this all my life as I have always felt that we could have been great friends as well as father and son. According to my mother, he had been seriously ill for some time with influenza, but had recovered and intended to go to work Monday morning. Sunday evening, they had gone to church, and on the way home he was eating cherries and flipping the pits at Mother, so he must have been feeling good. However, he did not awaken Monday morning.

As I grew older and heard people speak of my father, I found that he was loved and respected by all who knew him. This was made very clear to me when I returned to England fifty years later, accompanied by my wife, to attend the Coronation of Queen Elizabeth. At twenty-three, when he died, he had a very promising future, and commanded the admiration and respect of his contemporaries, not only for his work but as a man.

As a boy I was not too interested in family background but, upon my return, discussing it with my Uncle John was most interesting. He went with us to visit my old home which had been in the family for the past three hundred years. It looked like it was good for another three hundred, and was occupied by one of my cousins. The house was located in a beautiful basin-like valley, most of the surrounding hills having a fairly heavy growth of larch trees, and the valley having a luxuriant growth of various grasses belly deep to the few cattle and horses grazing there.

Being used to the Western ranges, I was quite surprised to see so few stock as the valley could have supported a much greater number. When I mentioned this to my uncle, he said it was because of conservation. However, it seemed to me they were carrying conservation a little too far.

The house was just as I remembered it, except that the spring from which we obtained our water had been improved with a concrete well and plank cover. The house was built of stone, with walls about eighteen inches thick, and even the roof shingles were of split rock, so no wonder it had lasted so many years. The parlor and bedrooms were being modernized with new ceramic fireplaces that had recently been developed. The large kitchen looked just as when I left, with the huge fireplace equipped with swinging irons on which one could hang a full-grown steer. The dairy also looked just the same—white flagstone floor, whitewashed walls and ceilings, and the wooden pegs on which we hung whole sides of bacon that Mother called "pictures."

Around the hills could be seen occasional open spaces with houses, all occupied by my cousins. In contrast to the older houses, one of my cousins, who had made sev-

eral trips to the United States and Australia, had built
a new home that was as modern as anything we have,
swimming pool and all. We would have enjoyed staying much longer but time was limited. We had to join
our party at Harwich to take the boat for The Hague,
Netherlands.

My father's death left Mother with three children,
the youngest being me. I was taken by an aunt for
some time, so that Mother could become adjusted. My
mother, in her way, was a remarkable woman. She was
a strict disciplinarian and tolerated no disobedience from
any of us. She had a good command of the English language and used words as an expert swordsman used a
rapier. I have seen her look a person in the eye and, with
a smile on her face, literally cut him to pieces, her language always precise and correct. In retrospect, she
reminds me of the English gentleman who above all had
to protect and preserve the King's English. It seems his
wife returned from a party earlier than expected, and
found him in a compromising position with one of the
maids. "Why, John!" she exclaimed, "I am surprised!"
Whereupon he replied, "No, my dear, you are astonished, I am surprised."

In all fairness I must not leave her after exposing just
the severe side of her personality. She was a charming
person, an accomplished pianist with an excellent so-
prano voice, and was very much in demand at the various social functions. She outlived two husbands, and
had been married about fifteen years to the third when
she died at eighty-four.

Education in England was mandatory from five to

fourteen, or to the completion of the eighth grade, but there was nothing to stop one starting earlier so at the age of three, I began my schooling. Because of my mother's coaching, I had little difficulty in keeping up with the older children and from the start I liked to study. Then, too, I was fortunate in having a good teacher. He, like my mother, was a disciplinarian, very strict, but very good.

During my sixth year Mother remarried. Though my stepfather was a good man, and always maintained a good home for us, I had a very definite feeling that since he was not my father I should not be a burden to him, so at that early age I started to make plans of my own. Our English school system was somewhat different from that of the United States. The school furnished everything; we took nothing into, or out of, the school, and had no homework. We were, however, required to work hard and concentrate while in school, and our examinations were conducted by a State Examiner who came in for that purpose, the teachers having no voice in the matter whatsoever.

With no homework, my evenings were free and in my ninth year, I went to work in a barbershop evenings and Saturdays. Here, too, the system was different from that of the United States, the barbers being required to cut hair and shave only, while a boy lathered the faces and prepared them for the barber. He also turned a crank to furnish power for a rotary brush used by the barbers. This equipment consisted of a line shaft over the chairs and a belt-driven brush similar in size and shape to a pastry rolling pin, and held in the same manner. When rotated at high speed this gave the hair a very brisk brushing. So I became busy with the pot and brush, lathering faces and sweeping up the shop.

Halfway through my eleventh year, I was graduated from the eighth grade, and, since there was no job available at the moment, I went to visit my mother's older sister and her husband. Having no children of their own, they had always been kind to me so I enjoyed being with them. He was a veterinarian with a good practice, together with a well-stocked farm of about one hundred and fifty acres. Since they had spent some years in the United States, I was naturally anxious to hear of their experiences, which were most interesting. It was from these discussions and their books on the West that I became imbued with the idea of one day going to America.

With this in mind, a study of horses, cattle, and sheep, became necessary, so I decided to obtain work along these lines. About this time there appeared, in the local paper, an advertisement for a carter's helper, and as the farm was about twenty miles away I rode my bike over to apply. It was a big farm with about sixty head of milk cows which supplied a large dairy in town, twelve huge workhorses, and a stable of hunters, jumpers and show horses. The carter and his helper cared for the workhorses and the trainer and his helper for the hunters, jumpers, and show horses.

I was finally ushered in for an interview, and as it was my first attempt to find employment, I was, naturally, quite nervous. The man for whom I hoped to work was quite large, and, I don't believe he was too impressed with my appearance. His first remark was, "Aren't you quite young to be starting to work?"

I replied, "I have finished my eighth grade, sir, and I am quite strong." However, I did not mention my age. We discussed in detail the work involved, but I still had some misgivings as to whether or not he would hire me. Finally, however, I was hired, and required to start on the first of the month which was three days away. I lived in (as did all the employees), and shared a room with the trainer's helper, Bill Smith, an expert horseman, who was my senior by several years. He was a small man as his job required, and had been at the farm for quite a long period of time. Our employer, Thomas Willoughby, was an avid huntsman and steeple-chaser. He loved to hunt so much that he always took Bill along for the first half of the hunt, and then changed horses, which explains the reason for a small man, so that his horse would be comparatively fresh at the half-way point. Here Bill would return directly to the stable with the tired horse, leaving his mount for the boss to finish the hunt.

About the fourth day on the job, I accompanied the carter to the city of Gloucester to get oil cake for the cows. Four huge horses were hooked up, tandem, to a very large wagon, and were handled by voice only—there were no reins of any sort. It was remarkable to see how those horses responded to the spoken command. We pulled up right alongside of the ship which was tied up to the pier. While in Gloucester, I went to the cathedral where, according to English history, Oliver

Cromwell stabled his horses and barricaded the building with bales of cotton.

My roommate, Bill Smith, turned out to be a pleasant chap, and, being much older and more experienced, was a great help to me. I enjoyed my work and got along well with my immediate superior, so life was quite pleasant.

For entertainment I found that Bill had an arrangement with the manager of one of the local theatres whereby he would place six placards in store windows in return for a ticket to the show. He introduced me and made similar arrangements, so, in this way, we saw all the best shows at no expense to us.

About a month after I came on the job, Bill informed me that he had a chance for a better place with more salary and was giving the usual thirty days' notice. Knowing I loved the hunters and jumpers and his work with them, he suggested that I apply for his job, but since I had never ridden a horse over a jump I was afraid to do so. Being the good scout he was, Bill said, "If I can get a suitable replacement, maybe I can get away with two weeks' notice, and I can teach you to ride those jumpers in two weeks." So, unknown to anyone but ourselves, we were up at three o'clock every morning taking my training. I'll have to admit that the first three or four jumps I was over before my horse but Bill was a good trainer, so two weeks and numerous bruises later, I was doing a passable job of riding. We both went to see our employer, Mr. Willoughby, Bill to give his notice and I to ask for his place. Mr. Willoughby was a little disturbed by the short notice and in a somewhat

severe tone of voice asked me if I could ride the jumpers.
I said, "I think so, sir." Turning to Bill, he ordered him
to have three horses ready next morning so that he could
judge for himself. I did not sleep well that night, knowing I would be riding under pressure the next day, and
was, to say the least, a bit nervous. The following morning we started across country, knowing that every move
was under observation. However, thanks again to Bill's
kindness and training, I did quite well, and, when it was
over, Mr. Willoughby said, "You'll do. Then, too, you
are considerably lighter than Bill, which suits my purpose very well."

As soon as I was established in my new job, I concentrated on learning all I could about horses in general,
and hunters, jumpers, and show horses in particular.
During this era, England was a nation of horse lovers, so
most of the wealthy people maintained stables of hunters, jumpers, and show horses. Even the tradespeople
with their delivery ponies had horse shows in that
classification.

Mr. Willoughby was an excellent horseman and
taught me many things, including the handling of a
horse at the water jump which is very hazardous, as
it consists of a brush fence three feet wide by six feet
high with a ditch eighteen feet wide filled with water
behind it, making a total width of twenty-one feet.
Under these conditions complete coordination between
horse and rider is required, so one rides with confidence,
which the horse seems to sense, and a snug—not too tight
or loose—rein.

Unfortunately for my boss, he was, as I have already

said, quite heavy. This was somewhat of a handicap for a horseman, but it made it possible for me to ride much more than I would otherwise have done. Soon I was riding with him on his private steeplechase track which he maintained for training purposes. He also maintained training facilities for the show horses, as their training was quite different, requiring showmanship instead of speed. We used to fasten lead weights on their front feet for about a week before the show, taking them off before showtime. This, of course, caused the horses to step unusually high, so they looked very good.

As soon as I was proficient, Mr. Willoughby started using me in races when regulations permitted. In the flat races, my weight of one hundred and five pounds was, of course, an asset. However, in the steeplechase, I had to carry lead as most races required that the horse carry eleven stone, which in America would be 154 pounds, still much less than my employer's 210 pounds.

He also used me in the horse shows with the jumpers and other show horses, and every time I rode he gave me an extra five-pound note, which at the rate of exchange then was twenty-four dollars. This boosted my income considerably, making it possible for me to go to the United States sooner than I had anticipated. This made me very happy as I was most anxious to get started. I had read many books on the West, including those about Buffalo Bill. My great ambition in those days was to be able to pick up a hat off the ground from the back of a running horse, which I later did without too much trouble.

My mother knew of my plans, and she approved as

she hoped one day to go, too, so thought I might be able to pave the way. Upon investigating the various ways to go, we decided it would be second class, as we had heard that the third-class, or steerage, passengers were treated more or less like cattle. This we did not want so decided it would be better to work a little longer in order to go second class. My mother had been saving my money for me and because of my age had to make the arrangements for my passage, which she did during November of 1902.

At this time I informed my employer of my plans and took the opportunity to tell him how much I appreciated his interest in me and the things he had taught me. On Christmas Eve, all the employees, comprising the butler, cook, scullery maid, housemaid, lady's maid, and carter's helper, all of whom lived on the premises, gave me a going-away party and they were all wonderful to me. We were served the typical English roast beef, together with several vegetables, mashed potatoes, and Christmas pudding flamed with brandy. Holly and mistletoe were hanging in great profusion, and the girls were enjoying my embarrassment every time they caught me under a bunch of mistletoe. It was unfortunate that I was too young to enjoy it. The boss and his lady joined us for a while and gave me a going-away present, together with very nice references, the main part of which I have forgotten. I remember he finished by saying, "And I wish him success in his future career."

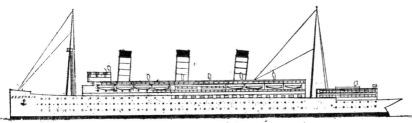

WHITE STAR LINER S.S. CYMRIC
The ship on which George E. Franklin came to America, as drawn
from memory by the author.

PLATE II

CHAPTER II

Westward Bound

CHRISTMAS DAY I WENT HOME TO BE WITH MY FOLKS during Christmas and the twenty-sixth. On the twenty-seventh, I traveled by train to Liverpool, and on the twenty-eighth, at 11:00 A.M., boarded the White Star Liner S.S. *Cymric*. As the time for my departure drew near, I thought Mother seemed quite nervous. I tried to convince her that I had been away from home a little over one year, worked among highly sophisticated people, and believed I had learned to take care of myself. This seemed to allay her fears and, as the ship pulled away, she was waving and smiling.

After pulling away from the pier, and eating lunch, we came on deck to see the sights. Traveling west through the Irish Sea, we passed Hollyhead and turned south through St. George's Channel to heave to off Queenstown, Ireland, where a tender brought more mail, passengers, and cargo. From there we went around the southerly tip of Ireland, and out into the Atlantic. We had many interesting people aboard. The food and services were excellent, and there were many games to play, as well as dancing, which I had not yet learned. One day, while walking around the deck, I found myself pretty well aft, and, as I looked over the rail, down into the third-class quarters, I was glad my mother and I had decided on second class. In mid-Atlantic we celebrated New Year's, arriving at Boston on January 6, 1903.

The crossing had been quite pleasant. I had shared my cabin with two very nice young men whom I had never met, but we were assigned the same cabin. One of the young men was Scotch, the other Irish, and, since I was English, Great Britain was well represented. My Scottish friend, Sandy McPherson, was the typical dour Scot, but I always suspected that Pat O'Brien, in addition to his natural Irish wit, most always had his tongue in cheek. As we left the ship at Boston and were walking along the pier toward the bow, Pat slowed down, and was very intently looking up at the anchor, until both Sandy and I became impatient. Sandy inquired as to his interest in the anchor, and Pat, in his typical Irish brogue, replied, "Shure and oid loike to see the mon thot handles thot pick."

Leaving Boston, we traveled by train west to Buffalo, and, since we all thought we might never come this way again, stayed there long enough to see Niagara Falls. At Des Moines, Iowa, Pat left us, and went north to Minneapolis, Minnesota, where he had relatives. Sandy continued on to Rawlins, Wyoming, where he also had relatives who were in the cattle business. As for me, I had no relatives and no particular place to go, so I decided to go on to Pocatello, Idaho, which sounded both Indian and interesting as it was close to the Fort Hall Reservation.

It was here that I really saw my first Indians. I had

gone back to the railroad car to get a package one of the ladies had forgotten, and, as I passed through the vestibule, there were two Indian braves in full regalia, war bonnets and all. They were quite tall, and as they stood erect, with their arms folded across their chests, looked quite formidable. I breathed a sigh of relief when I had passed them. After all, these were my first Indians, and I hardly knew what to expect.

After an overnight stay in Pocatello, which was a good-sized town and railroad terminal, I decided that the range country was what I really wanted and traveled on to American Falls, Idaho, which was open enough range country for anyone. When I first saw it, the town comprised the Fall Creek Store, the Oliver House, and, of course, the inevitable saloon where the stockmen for miles around congregated.

The Oliver House was originally a log building to which several log additions had been added, making about five rooms available for rent. The day I arrived, there was none available, so Charles Howard, manager of the store, kindly offered to share his bed, which was in the store building. Next morning, as we went to the Oliver House for breakfast, we realized that everyone seemed tense and excited. We followed the crowd to the railroad bridge and there I froze in shocked amazement and wished I were back in England. It seems the cattlemen had rounded up some sheepherders overnight and hanged them. I then learned that a war existed between the cattlemen and the sheepmen. The cattlemen were violently opposed to the intrusion of sheep and determined to keep them out. This war continued for some time and was bitterly fought by both sides until it reached a climax about 1907 when an ex-cattleman brought a large herd of sheep into the Big Horn Basin of Wyoming. The result of a vicious attack on the camp of a former cattleman was an investigation that led to the peaceful arbitration of range disputes, and the locating of sheep and cattle on different sections of the range, supervised by the United States government.

Because of my shocking experience during the early morning, I was quite upset. However, here I was, in strange country, among strange people, running out of money, and no job. I approached a kindly-looking rancher, who was, from his speech, obviously of Scandinavian extraction, and asked about work. He looked me over rather skeptically, I thought, and informed me that this was cattle country and there was no work during the winter months, except feeding, and there was more than enough help for that. I was about to leave and look elsewhere when he said, "If you like, you can come to the ranch and do chores for your board until spring when there will be work to do." I was very glad to accept his kind offer and he told me to be ready early next morning as the ranch was about twenty miles distant. I had not unpacked my trunk, which was quite large, as I had brought a good supply of clothing including a pair of English riding boots. These boots evoked a lot of laughter at my expense, but I continued to wear them as they were much more comfortable than the high-heeled Western boot.

CHAPTER III

Becoming Americanized

SINCE THE INCIDENTS RELATED HERE COVER A LARGE territory, comprising areas in five states (Idaho, Nevada, Utah, Wyoming, and Montana), it might be of interest to give a brief description of the terrain.

I must admit that, as I first beheld this vast country with the snow-clad mountains in the distance, I was homesick for my old home in England. But in the months and years that followed, as I rode, sometimes from sunup to sundown, without seeing a fence or a human habitation, I learned to love those wide horizons with their deserts, valleys, and mountains, which were awe-inspiring because of their tremendous dimensions as well as their majestic beauty. I remember one day standing on a high mountain with the dense white clouds swirling at my feet far as the eye could see, with an occasional peak pushing up through in the distance, and the vast blue dome of the sky above. I was up there all alone, and I felt very small, very insignificant, and very close to my maker. Many years later a song was written that expresses my thoughts that day much more adequately than I could ever do. With apologies to the authors, Florence Tarr and Fay Foster, I repeat a few lines for you: "Oh, the place where I worship is the wide open spaces, built by the hand of the Lord. Where the trees of the forest are like pipes of an organ and the breeze plays an amen chord. Oh, the stars are the candles and they light up the mountains, mountains are altars of God."

American Falls, Idaho, was an early campsite on the old Oregon Trail, and twenty miles downriver is a grim reminder of the wagon-train days called Massacre Rocks. Why the wagon master of this train ever decided to camp in this particular spot is beyond my comprehension as it was so obviously a trap from which no one could escape. There was a vertical cliff about seventy feet in height and semicircular in shape, extending from the river on the east and running in an unbroken line almost to the river on the west, making it necessary to blast a road at the eastern end in order to get wagons down to the lower level. This, no doubt, appeared to the train master as a very desirable little amphitheater in which to camp, but he obviously gave no thought to defense or escape, and, apparently, posted no guards to warn them of danger. They circled their wagons that tragic night, perfect prey for the Indians who slaughtered them from the cliffs above where they were protected and comparatively safe. Many of the irons from the burned wagons were still there when I first saw the place in 1903. Unfortunately, many of our historical monuments are not placed until so long after the incident happened that they are incorrectly located. This one is about a mile downstream from where the event actually happened.

American Falls is now a thriving town of about

twenty-two hundred people and is the center of a huge irrigation project which allows thousands of acres to be cultivated. The American Falls Dam across the Snake River is nearly a mile long, and creates a lake thirty-six miles in length, with a storage capacity of 1,700,000 acre-feet.

I met the friendly rancher, John Eliason, early next morning and we started for the ranch. Upon arriving there, I found that he had four grown sons, the youngest being sixteen, and three daughters. They attended one session of school each day from 10:00 A.M. to 3:00 P.M. This arrangement made it possible for them to feed the cattle, of which there were about four hundred, mornings and evenings. I may have been young and inexperienced, but I was not naïve, and it was obvious that Mr. Eliason was giving me a home until spring, which I did not like, as I wanted to pay my way or at least earn my board and room. This opportunity presented itself when I found they contemplated quite a building program consisting of a huge barn and several smaller buildings. They had the necessary carpenter tools but knew very little about reading the steel square, so most everything was being done the hard way. My stepfather was superintendent of a large wood turning mill in England, and was also an excellent carpenter and cabinetmaker. From him I had learned, at least in an elementary way, to read the steel square so that I could lay out rafters, joists, and the necessary component parts of a building. This speeded up their work and made them all very happy. As for me, I felt that I was at least paying my way. This little episode opened other doors for me as the family seemed to realize that my education, even though it was only through eighth grade as taught in England, was far superior to the schools in the valley. As a result I was able to help the children with their studies. This soon became common knowledge and I was kept busy helping the children— and their parents with building problems. Now at last I felt at home with these wonderful people, and I became known as "that English kid that was as handy as a pocket in a shirt."

I tried very hard to learn about ranching, irrigating, putting up hay, cleaning irrigation ditches, mending fences, getting in firewood for the winter, and other chores too numerous to mention that were necessary to the life of a rancher.

There were, of course, lighter moments such as the dances that were held weekly during the winter months. Even though some of the ranchers had to ride as far as thirty-five to forty miles, we usually had a good crowd and an enjoyable time. This was quite natural as that was the only amusement they had, and many of them had never known anything else. As for me, I missed the wonderful shows I had been accustomed to in England, but here I was in a new environment trying to adjust to a new life. Most everyone in the valley was of Scandinavian or Polish extraction, so the polka, schottische, square dance, and waltz were the popular dances of that era and it was necessary that I learn to dance them, which I did with the aid of one of the Eliason girls. Lots of refreshments were available as they usually danced until daylight.

During my first winter the jackrabbits were so numerous that they reached disaster proportions and a bounty of four cents per head was placed on them. We actually conducted huge drives like cattle roundups, built wire mesh corrals with wings extending from the entrance at an angle of forty-five degrees for quite some distance to trap them, and hunted them on horseback. I have seen as many as fifteen thousand killed in one drive.

So ended my first year in America. I have to admit that through the Christmas and New Year's holidays I experienced some nostalgia but I was settling down and rapidly becoming Americanized.

CHAPTER IV

Horse Roundup

DURING EARLY 1904 I WAS INVITED TO GO ALONG ON A horse roundup, and, even though I had ridden very little the past year, and then on gentle horses, I was most happy to come. No one had been told of my riding experiences in England, as I realized that everything was so different here, and I wanted to become familiar with the customs of the West. In England, riding horses were used almost exclusively for pleasure, so they were trained accordingly. We were taught to post, and, in training, our arms were strapped to our sides until we learned to hold them that way.

In the West, a cattleman's horses were the tools of his trade. He had never heard of posting, and if he had, he would have discarded it, since the long hours in the saddle would have made it most uncomfortable. He sat down in the saddle, and, to all intents and purposes, literally became a part of his horse. Horses were trained to be cutting horses for the purpose of cutting cattle out of the herd for branding, and for shipping beef in the fall, and when a rider mounted, he had better be prepared to ride because a cow horse could hit full stride in about two jumps, turn on a dime, and once he knew the animal you wanted you could forget your horse because he usually knew his business. Horses were also trained to rope from, and once you had your lasso on an animal and the dallies around the horn, the horse always maintained the proper position relative to the lassoed animal; otherwise, horse, and rider, might be in trouble. Cow horses were also trained to neck-rein, and, with a well-trained horse, the slightest touch of the rein would cause him to turn. The cowboy is a very busy person when working cattle as he has to handle his rope, maintain contact with the animal he wants, and ride, so neck-reining is very necessary. It is quite different from riding in England where you use both hands and concentrate on your horse. One thing that really amazed me when I first saw it was the way a cowboy could dismount, drop his bridle reins, and walk away leaving his horse ground hitched, as they called it. Naturally, I was curious as to how a horse was trained to do this, and, later, I learned that usually when a horse is first ridden a hackamore is employed. The hackamore is a braided rawhide noseband with the customary leather band over the head, and a very strong, usually bell cord, double rope, attached to the noseband under the jaw by a special knot with a loop below, and the other end of the rope over the neck, paralleling the leather band. This makes a very strong arrangement and it is used almost exclusively in the breaking of bronchos. When a bridle is first used the cowboy deliberately puts in a very severe Spanish spade bit with a long spade extending well back into the mouth. A horse, when attempting to leave, steps on the reins, driving the spade against the roof of the mouth with very painful results. This device quickly breaks a horse to stand where you leave him.

It was with mixed emotions that I awaited the start of the roundup, as it was surely going to be a new experience for me, and another facet of the rancher's life.

During roundup time, either for horses or cattle, several of the ranchers combined their forces and worked together, using the corrals at the ranch nearest to their work. These corrals were always, or most always, very strong circular enclosures with snubbing posts at the center.

We started at one end of the territory to be covered with the usual chuck wagon and remuda, as we sometimes changed horses several times each day. I felt very inadequate with all those experienced riders, as I had not yet seen a bucking horse, and very little roping, but in the days to come I would be seeing plenty of both. We usually rode in pairs, radiating out from the corral to be used like spokes in a wheel until we reached the outer limits of the range to be covered. There we started rounding up the horses and heading them toward the corral. This was worth all the days of waiting, and I was glad at this moment that I had not mentioned my days in England since I did not know how I was going to do in this new environment with everything so different—the horses, equipment, even the terrain over which we were riding. When we reached the limits of the range allotted to us, and started gathering horses, I forgot all my fears and inhibitions. I was riding again.

The terrain was strange to me as were the various types of brush and vegetation, and, because of my inexperience, I jumped my horse into a bunch of chapparral, causing a dirty spill. Fortunately, neither of us was badly hurt. My riding partner roped my horse for me, so I mounted and continued the ride.

That night, after we had eaten and were sitting around the campfire swapping yarns, one of the older men suddenly said, "Say, Kid, where did you learn to ride? No green English kid could ride that way first time out." I took this as a compliment, since it came from a man recognized as one of the best riders in the country. Feeling that I was now accepted by the group, I told them of my experiences in fox hunting, steeple-chasing, horse shows, and such. When I had finished, he said, "I knew you had been around the way you handled your horse."

It was during these evenings around the campfire that I learned that certain quite harmless words and phrases, used in England, had an entirely different connotation when used in the United States. This was a source of amusement to the boys, but of embarrassment to me. One typical American expression that meant nothing to me, but almost got me in trouble, was the well-known, and often-used, S.O.B. It meant no more to me than son-of-a-gun, as I had not thought to analyze its implications. One day, after the boys had been having a little fun at my expense, I called one of them an S.O.B. Lars Nelson, a huge Dane, who had been an able seaman in his younger days, and had been in many of the world's ports, grabbed me by the front of my shirt and held me at arm's length. "What did you say?" he exclaimed. I repeated it, and he asked me what the fighting word was in the country I came from. I told him and he replied, "Well, brother, what you just said is the fighting word here, and when you say it you had better smile." Then they all had a good hearty laugh at my expense, but I was learning.

The days that followed were just as exciting as the first, and we covered a lot of territory, changing corrals every three or four days. Every weekend we ran in a bunch of horses and pulled straws to see who rode those that were picked. I had never ridden a broncho and sometimes wondered what would happen when my turn came, as it had to sooner or later when I pulled that short straw.

When my time finally came, I drew a little buckskin with a black stripe down his back and a black tail. The boys got him saddled and, with some misgivings as to my ability, I mounted and away we went. We usually rode a horse the first time in a circular corral, already mentioned, so that, if the rider was thrown, it wouldn't be necessary to round up the horse. Riding in an enclosure this small was new and strange to me and every time the horse bucked close to the fence I thought my leg would be crushed, and pulled my foot out of the stirrup, throwing myself off balance, so the next jump he threw me. After this happened several times, I told the boys I thought I could ride him outside if they would keep him out of the fences. He was, fortunately for me, not a hard horse to ride. He went high, wide, and handsome, but straight ahead, no swapping ends, sunfishing or other tricks bronchos usually do. Since, in my steeplechasing days, I had been used to riding high and wide, this did not bother me too much, and I soon completed my first ride as a bronchobuster.

Toward the end of the roundup, we were seated around the campfire swapping yarns one evening when the stories turned to the ghost variety which were popular at that time, and one of the cowboys said, "Hell, I'm not afraid of ghosts. It's the live ones that bother me." This, of course, planted an idea in some fertile mind and the fun began. About that time the forerunner of the dry farmers had filed on a homestead near the ranch where we expected to spend the weekend, and two of the boys rode over to see if he would cooperate in a little joke on our fearless compatriot. He thought it was a good joke, and agreed to play dead for the occasion. So we informed our friend that the homesteader had died, and, since he was the only one not afraid of the dead, he was elected to sit overnight with him. He took it in stride and we went ahead. We had arranged with the homesteader to raise up under the sheet, at the sound of a cat's meow, and scare our friend out of his wits.

Everything went according to plan. We all rode out and our man looked the situation over, then placed a chair beside the bed and started to read, totally unconcerned. We went outside to see the fun through the windows, but when the cat meowed, and the supposed corpse rose up under the sheet, the scene changed, and, instead of running for his life, our hero calmly drew his .45, tapped the would-be ghost lightly over the head, and said, "Lay down, you S.O.B., you're supposed to be dead!" We never did find out who the informer was who had turned the joke on the jokesters, but our intended victim maintained his reputation for invincibility intact.

The roundup over, the colts branded, and the male colts castrated, we returned to the routine maintenance work of the ranch, and, when this was done, it was time to put up hay, of which there were three crops each year. Most of the hay was alfalfa, or as it was called locally, lucerne. Usually the third or last crop was used for seed as there was a growing demand, at good prices. The method of putting up hay was entirely foreign to me as everything was done by hand in England. Here most of it was done by horsepower. It was raked by a horse-drawn rake, and stacked by a Jackson Fork that took a whole wagonload in about six fork loads. This was also operated by horsepower.

CHAPTER V

Beef Roundup

WITH THE HAYING DONE, IT WAS TIME FOR THE BEEF
roundup, which was somewhat different from the spring
roundup, when the calves, like the colts, were branded.
The remuda and chuck wagon were still a very necessary part of our equipment, as the horses were worked
very hard, and we needed several changes for each man.
The men also worked very hard and had to be fed.

The beef cattle were, of course, handled differently
than in the spring, as they were now fat and had to be
moved as slowly as possible, so that they would not lose
the weight they had acquired during the summer. We
gathered them as slowly as possible, and held them in
open country, always working them toward the shipping point and letting them graze on the way.

When the herd was gathered, the ranchers appointed
one of their neighbors, Bill Oliver, to act as trail boss.
He was an old-timer and certainly knew his business,
so he called us together to give us our positions. "You,
Bill and Jack, ride point, and head them toward the
falls. Jim, Harry, Fred, and Tom, take the swing, and
I'll take the kid with me and bring up the drag." The
remuda was handled by the wrangler, and the chuck
wagon by the cook. Both the chuck wagon and the

remuda were off to one side of the herd, traveling parallel to it to avoid the dust as much as possible. Up on the point of the herd, Bill and Jack kept direction. Along the flank, Jim, Harry, Fred, and Tom watched to see that no animal strayed. Behind them all rode the boss and me, constantly alert, keeping the drags going, and seeing that no steers turned back. Occasionally one of the boys from the swing would drop back to give us a hand. About eleven hundred big steers were on the march.

I had often wondered why cowboys wore their mufflers backwards with the knot at the back of the neck. Now, riding drag, I knew. It was so that they could wear them across the bridge of the nose, covering both nose and mouth, thereby serving as a filter to keep out the dust. In fairness to all the riders, the positions of point, swing, and drag were changed regularly, since it would be unfair to keep riders in the drag position at all times.

In due time we arrived at the shipping point at American Falls, loaded the steers, and separated, each going his own way. Some went with the steers to Chicago, where they ran wild with the ladies of easy virtue until their summer wages were gone, then, sadder but wiser, they returned to the ranches where they could at least eat and sleep until spring came again. Others went back to the ranches, and I went on to Idaho Falls to work the fall and winter run at the sugar factory, which gave me about four months' work not available on the ranches.

CHAPTER VI

Wild Horse Episode

IN THE EARLY SPRING OF 1905, A GROUP OF YOUNG MEN decided to go on a wild horse hunt. Horses were in demand throughout the East, and brought good prices. I had ridden with most of these fellows and I believed they were my friends, so I asked to go along. They informed me that this was no picnic and that there were a lot of hardships involved which they doubted I could take. However, after considerable discussion, and, I believe, due to the fact that I had about two hundred and fifty dollars to help buy the necessary supplies, they agreed to accept me as one of the group.

From the time I had started to work for Mr. Willoughby, I had become intensely interested in horses of all kinds. Before leaving England I had read considerably about the origin and history of the American mustang, so I was astonished at the number of people who believed that the horse was a native of the Americas. In fact, very few in this area thought or knew otherwise. The horse was, of course, brought to the Americas by Cortez in 1519 when he led his expedition for the conquest of Mexico. He landed at what is now Vera Cruz, and, with his conquistadores, brought a number of stallions and mares. Later, Alvarado joined Cortez and brought more horses. Still later, Narváez brought more. After his campaign, which lasted about two years, Cortez returned to Spain, and many of the horses were turned loose on their own. These horses multiplied rapidly and migrated through Mexico and as far west as California and the Pacific Northwest. Their offspring became known as mustangs. These mustangs were now being caught and broken by the Indians, and it was this that changed the Indians' whole way of life, because, being on horseback, they could more easily hunt the buffalo which was the main source of their existence, not only for food, but also for robes, tepees, and other things. Even the buffalo chips were used to burn. Then, too, the horse made them much more mobile, and they were able to travel greater distances in much less time. As a result of the acquisition of the mustangs by the Indians, they became some of the world's best riders.

During the years of indiscriminate inbreeding, the mustang had, of course, degenerated until they were quite small, but with the advent of the cattlemen, they started to improve the breed. In order to obtain saddle stock for their operations, they imported Thoroughbred, Morgan and Steeldust stallions. I was quite familiar with the Thoroughbreds, having worked with them in England, but knew very little about the Morgans and Steeldust breeds, since they were American products. Consequently, I did some research on authoritative reference data relative to these two breeds.

The Morgan is classified as being the first true American breed of horse. Start of the breed was from a dark bay with black legs, mane, and tail, foaled in Vermont in 1789, and owned by a man named Justin Morgan. This horse became known as Justin Morgan, a small horse of about fourteen hands and weighing about one thousand pounds. His breeding was unknown, but from his description he was no doubt of Arabian and Thoroughbred origin. All Morgans have bred true to the conformation, type, disposition, and style of Justin Morgan. Cattlemen have said its intelligence, tractability, surefootedness, strength, and gameness make the Morgan very popular. In fact, the Morgan has all the attributes the cowman looks for in his horse.

The Steeldust horses can trace their ancestry to a small chestnut Thoroughbred stallion imported from England to west Texas after the Civil War, and brought here by a man named Steele. The horse was originally named Dustistan, but Steele shortened his name to "Dusty," and later he was referred to as "Steele's Dust," then later to "Steeldust." He was about fourteen and a half hands high and weighed about eleven hundred pounds. One of the fastest short-distance runners, he seldom lost a race. He had the faculty of producing colts of his own color, conformation, disposition, and speed, and is credited with being the ancestor of the present-day quarterhorse. These three breeds were chiefly responsible for the western cow horse as we know him today. The progeny of the horses left by Cortez and his conquistadores, the mustangs, roamed over most of the West, usually back in the interior where few riders penetrated. Here, too, the breed was improving, due no doubt to the law of the survival of the fittest, since once a stallion had gained control of a harem of mares he tolerated no competition, so the weaker or older stallions were driven off, and quite often went to the wild bunch. The fight to maintain control was a continuing and vicious thing, and, once two range stallions engaged in battle, it seldom ended until one of them was completely vanquished and quite often killed. Consequently, even though they were not the largest or strongest of the domesticated herd, they were usually much larger and stronger than the mustang, and certainly better bred, which was reflected in the improvement of the mustang herd.

These, then, were the horses we proposed to hunt, and since we had decided to cover a large territory ranging from Promontory Point in the Great Salt Lake of Utah to the Big and Little Lost River country in Idaho, a lot of preparatory work was necessary.

As it was early spring and quite cold, we used a sheep wagon so that we could at least cook and eat inside. Then, too, we could carry quite a few tools which were necessary to our project, and sleep two of our crew. For the other two members of the crew, we built an extension to the back of the wagon which was quite comfortable. We started near the Lost River sinks, and built our first corral or trap there. This corral was one of our best, consisting of a blind box canyon making it necessary to fence one end only, as the other end of the canyon was a vertical wall at least one hundred feet in height. It was quite large, with a good stream of water and plenty of feed, so that a fairly large bunch of horses could be held there for some time. The approach was from a wide flat valley, and the horses did not realize they were being trapped until the valley narrowed and the walls became vertical, then it was too late. We built our gate and fence at a narrow point in the canyon, and continued south to establish other corrals or traps. Most of these were open-end box canyons requiring two fences, always built at the narrowest points. Occasionally we would make arrangement with a rancher, if he was not too far distant.

In due time we reached the area of Promontory Point, where we established our last corral. We were now ready for our drive north, and, we hoped, the corraling of many horses. With the completion of our preparations, it became necessary for some of us to go to the ranch and assemble a remuda consisting of at least twenty-four horses, as we planned to relay four times each day, and the horses would need a well-earned rest every third day to keep them in top form. We also had to get a supply of oats, as the horses would be working hard and need to be fed grain as well as the usual pasture, so two men went to the ranch for the horses and the others journeyed to Brigham City, Utah, to purchase oats and supplies.

We were necessarily quite methodical in our work, as we had a large area to cover and could not afford any errors in judgment. At each station from which we worked, we laid out a circle of about thirty miles, our corral or station being the center. By projecting a line through the periphery of the circle every forty-five degrees, we could establish landmarks to be used. We left the corral from which we were working early in the morning with four horses each, staking them out at approximately equal distances so that we could pick up our relays on the return drive to the corral. This not only gave us fresh horses, it gave us four well-broken horses that were being fed grain, which of course they liked, heading for the corral, and, at each relay, we had four more. This was a great help as the wild horses quite naturally followed them in. We tried to cover a forty-five degree section every day, requiring eight days at each station. When enough horses had been rounded up, we notified our buyer and arranged to meet him at the nearest shipping point. There were no negotiations as to price as our contract called for a flat fifteen dollars per head for all we could round up. So we gradually worked north until we came to our last camp in the Lost River country. This was an ideal camp with plenty of water and food. Then, too, we noticed that the horses seemed to be of better stock, probably because there were several horse ranches in the area where well-bred horses had been raised for some years, and there were always some losses to the wild bunch.

About the fourth day of our run at this camp, I noticed a black gelding, about sixteen and a half hands high and weighing about twelve hundred pounds, that was outstanding, and I wondered why they had castrated him as he would have been a magnificent stallion. Fortunately, whoever had performed the operation, had, in rangeland parlance, "cut him proud," and he still had the fire and spirit of a stallion. During my brief association with Thoroughbreds, I had seen many beautiful horses but none to compare with this one, and, as he stood on a little hill snorting defiance, with his mane flying in the wind, I determined that I must have him no matter what the cost. He reminded me of an article I had once read by Abd-El Kadar, Arab chieftain, horseman, scholar, and gentleman, which went as follows:

If, in the course of your life, you come upon a horse of noble origin, with large lively eyes, wide apart, and black broad nostrils, close together; whose neck, shoulders, haunches and buttocks are long, while his forehead, loins, flank and limbs are broad; with the back, the shinbone, the pasterns and the dock short; the whole accompanied by a soft skin, fine flexible hair, powerful respiratory organs, and good feet, with heels well off the ground, hasten to secure him, if you can induce the owner to sell, and return thanks to Allah morning and night for having sent thou a blessing.

This was such a horse. I had assumed that this horse was one of the wild bunch, but on closer observation I noticed that he carried a very small brand which placed him in another category, and, upon investigation, I found that the ranch was about fifty miles distant, and owned by a man named George Stanger who had been raising fine horses for the past twenty years. I wrote him immediately telling him that I had the horse, and asked him to quote a price, as I would like to buy him. His reply was definitely not what I expected, and left me with mixed emotions to say the least. It follows:

DEAR MR. FRANKLIN:

Your letter relative to a horse wearing my brand has been received. This horse is, I believe, one that disappeared from our range about two years ago. He was ridden a few times as a three year old and then threw a new man I had hired. This man tried a second time, and was not only thrown again, but would have been mangled or killed had not one of the boys jumped his horse between them. Since that time he has not been ridden and is considered an incorrigible outlaw.

He does have a good background being out of a well bred black mare and sired by a thoroughbred stallion I imported from England.

If you can break him to ride, which I seriously doubt, you may have him and consider this letter as your bill of sale or certificate of ownership as you wish.

Sincerely yours,
GEORGE STANGER

After reading this letter, which indicated a person of intelligence and education, I had a desire to meet him, and intended to do so in the not too distant future. In the meantime I had to decide what to do about this horse.

In the days that followed I studied him every chance I had, trying to determine what made him go bad, since he had been ridden a few times with normal results. I finally came to the conclusion that this particular rider had abused him so badly, that, being the spirited horse he was, he had finally rebelled and turned savage. If this diagnosis was correct, I felt that, with patience and proper handling, he could be reclaimed.

This being our last and best station, not only because our enclosure was larger but because the horses were of better quality, we decided to spend more time here, so I used it to good advantage, riding among the herd every day when we came in from the run, until they were used to having me around. I finally started roping him out of the herd every afternoon, and working with him. At first he fought like a tiger, but each day he became a little more docile, and I was encouraged with our progress. All through the drive we had problems with horses trying to break away, especially when they felt they were being trapped. These we attempted to crease, which is shooting a horse at the crest of the neck, knocking him out temporarily, and when he recovers he is usually much more docile. Creasing requires extremely accurate marksmanship, because a low shot would sever the spinal cord, resulting in a dead horse. Some of the horses were too valuable to take this gamble, so we used other methods such as roping and the like. Occasionally one would get away, and I remember one we called the Gray Ghost, a dapple gray stallion, weighing about twelve hundred pounds. He was very fast, and, when he felt he was being trapped, he would go over you, if he could not go around, as one of the boys found out when he charged, knocking his horse off his feet. He was still running free when we left. We were now getting near the end of our drive, and, by persistent and patient work, I finally was able to get a hackamore on my horse without quite so much resistance on his part. Because of his coal-black color I had named him "Midnight."

When the drive was completed and we had our final accounting, we found that we had corraled and sold slightly more than six hundred head at fifteen dollars per head. When all expenses were paid, we each had about two thousand dollars so we were all very happy with our exceptionally good luck. The boys returned one hundred dollars of the two hundred and fifty I had advanced, to equalize our participation in the venture, and I gave them the hundred for Midnight. We had, after all, found him in the wild bunch.

The boys went back to the ranches from where we had started, but I had already made arrangements with the nearest rancher for the use of a bunk, stall, and circle corral, as I still had a horse to ride.

As soon as we were established in our quarters at the ranch, I started to concentrate on the task before me. I had no illusions about this horse, and knew he was an outlaw as well as a potential man-killer. However, I felt that by perseverance, patience, and kindness, we had already come a long way toward dispelling the fear and hatred he had for a man, and, since time was relatively unimportant, there was no use trying to hurry, and, possibly, destroy everything that had been done. With this in mind, I proceeded very cautiously, working with him in the circle corral where I had a snubbing post to use. After several days he lost some of his nervousness, and I felt that I could work around him more freely, and that the time for my big battle was at hand. It was beneath the dignity of a top Broncho Buster or Contest Rider to use any artificial means whatsoever to help him ride. He rode free and easy, and raked his horse from shoulder to flank. However, I was not by any stretch of the imagination a top Broncho Buster and had no reputation as such to maintain, so I intended to use any means at my disposal in order to ride. I had deliberately chosen to do this alone, since I did not want a lot of excitement, and, knowing this horse's background, I had to ride him or face the possibility of being trampled, as I did not yet know how he would react under a saddle. Several devices to help me ride passed through my mind, including the hobbling of my stirrups, but I finally went to Idaho Falls and had the blacksmith make me a pair of spurs with a definite turned-up hook on the shank above the rowel which I could hook in the strands of my cinch, then I purchased an extra-strong cinch, and after checking all my equipment, decided I was ready to go. Early next morning, I snubbed him up to the snubbing post with a knot that could be easily loosened from the saddle, and, with a prayer on my lips, mounted, and made sure that my spurs were securely hooked, because, this was no fifteen-second rodeo ride, with pick-up men to take you off. It was to a finish.

When I turned him loose, he stood for a moment as if undecided, then dropped his head and away we went. Every time he hit the ground it felt as though a powerful man was driving a fist into my solar plexus, but my spurs held, and, after what seemed an eternity, he stopped, completely bucked out, and stood with his head down, feet spread apart, and trembling like a leaf. As for me, I was gulping large mouthfuls of air as he had almost completely knocked the wind out of me, and, if he had continued a little longer, I would have fallen from sheer helplessness. As soon as I had regained control of myself, I reached over, patted him on the neck, and talked to him quietly. He had not been spurred or abused in any way, and I believe he understood. I was not proud, satisfied, perhaps, with my accomplishment. True, I had ridden him, but I had taken an unfair advantage without which I could never have done it. Now that I had ridden him to a satisfactory conclusion, and felt that the worst was behind us, I had time to ride him on short jaunts, and, as the days passed, he became much easier to handle, and, knowing most horses weakness for sugar, I bought a package of the lump variety for him, which he eagerly ate without hesitation. He was now accustomed to the bridle, and I felt, gentle enough to be shod, as he had to be, since I had some long trips ahead for him.

Unlike most young men of those days, I did not smoke, drink, or gamble, for which I can take no credit since I had no desire for those things, so there was no incentive. I did, however, have a fondness for beautiful things, such as clothes, and since I was now the owner of one of the most beautiful and outstanding horses I had ever seen, and had a bank account of about two thousand dollars, I was determined that my riding equipment should complement the horse. I had heard of a saddler down Nevada way named Garcia, who was renowned for his work all over the West, and, as I was footloose and fancy free, I decided to drift down that way. I had gradually been hardening Midnight by feeding him grain, and by exercising him every day, so that he was now in excellent shape, shod, and ready for the road.

CHAPTER VII

I Go To See Garcia

ON MY WAY SOUTH I TOOK A LITTLE SIDE TRIP TO VISIT
my benefactor, Mr. Stanger. He was astonished when
I rode in on the "incorrigible outlaw," but made me
very welcome, and I discovered that he, too, had come
from England, loved horses, and enjoyed raising them.
He seemed quite anxious to learn how I had trained this
horse, especially because of my age. I told him that my
training in England, even though of short duration, was
intensive, and all with Thoroughbreds, and that Midnight had been handled as a Thoroughbred instead of as a
mustang, which I was sure he would not have tolerated.
I mentioned the small size of his brand and he informed
me that he used as small a brand as possible, so as not
to disfigure the fine animals he raised. From this we
drifted into a discussion of brands that was most interesting. He told me a story of a wealthy Bostonian,
owner of a big ranch which he seldom saw more than
once each year. It seems the rustlers were working on
his herds, and he thought he would use a little psychol-
ogy, so he registered a brand "ICU" and branded all
his stock. "Now let's see them use a running iron on
that," he said, and left for Boston. The following year,
as he was riding with his foreman, he noticed quite a

few cattle carrying an "ICU2" brand. I then told him
of a friend of mine with a sense of humor, who had
registered a brand, "Two Lazy Two Pee." He then said,
"Since we are discussing brands, I would like you to stay
over a few days and meet a friend of mine who has a very
unique brand. He is an old Spanish grandee, Gaudalupe
Valdez, who came to this country many years ago and
raises fine horses."

Early next morning we started for the Valdez Ranch,
which was about thirty miles distant, and, upon arrival were greeted very hospitably, as was the custom
in those days. Stanger and Valdez were apparently old
friends, so we had a most enjoyable visit, and Stanger
said, "My young friend here is interested in unusual
brands, and I thought he might like to see yours." The
following morning I was to see the most unusual means
of identification. It was not a brand in the true sense
of the word. Señor Valdez raised beautiful horses, and
he did not want to disfigure them in any way, so he
just made a small incision under the mane and inserted
a dime, which was hardly visible when it healed, but
could be felt, and was a positive means of identification.

We spent two days at the Valdez Ranch, which was
furnished in a definite Spanish motif I loved, as I did
most things Spanish—music, dancing, architecture, yes,
and their señoritas. Our visit over, I bid Stanger and
Valdez good-bye and started south, crossing the Snake
River Basin to Minidoka, then on to Rupert and Burley.
Here, after a good rest and stocking my saddlebags, including a tobacco can filled with a paste made of wood
ashes and kerosene, which I always carried to enable

me to start a fire wherever we camped overnight, even though things might be quite damp, we left the Snake River and followed Goose Creek southwest through the Sawtooth Mountains, and on to San Jacinto in northeastern Nevada.

This was cow country and was quite different from Idaho on the other side of the mountains. There, the dry land wheat farmers were slowly but surely getting established, and, while the ranchers did not like it, there was nothing they could do to prevent it. Here in Nevada there was no farming; the ranchers were interested in raising beef cattle, even to the point that they used canned milk. The lowlands or meadows were sown to timothy and clover and very little, if any, of it was plowed. The timothy and clover were sown into the natural wild grasses, and the combination made an excellent hay of which one crop was stacked each year, instead of the three crops of alfalfa as in Idaho. The method of putting it up was also much different. Buck rakes were generally used, and these consisted of an arrangement of teeth about ten feet long, equally spaced for a width of about ten feet, and pushed, not pulled, as is usually the case. The buck rake bucked the hay onto a huge fork which in turn elevated it to the stack. As work horses in this area were used only for haying and the hauling in of supplies, which was usually done in the fall, they only worked about three months of the year. As a result they were seldom more than half broken, so we had a lively time when haying started until they had worked a few days and settled down.

Even the language and equipment were different. A

rawhide reata replaced the hard-twist lasso, the hackamore became a jaquima, a maverick became an orejano,
and hair ropes, which I had never before seen, were com-
monplace, being used both as ornamentation and to sur-
round one's bed at night to keep the rattlesnakes out.
They would not cross a hair rope. Many of the boys
were also wearing tapaderas, which were seldom seen in
Idaho.

These Spanish words and customs were no doubt
brought to Nevada by the early Californios who had
drifted in many years before. Then, too, there were
many Spanish Basque ranchers and Mexicans who had
maintained the Spanish language. Another thing of
interest to me was a breed of horse I had never before
seen, the Palomino. He was a beautiful horse, the color
of new gold, with white mane and tail. By research I
found that he had been known as the "Dorado" and
the "Isabella," the latter because Queen Isabella of
Spain had a stable of these golden horses.

During Spain's conquest of the new world, she presented one of these golden stallions to a conquistador
named Juan Del Palomino, and it is said that after the
conquest of Mexico he stayed on and bred these golden
horses. They became known as the horses of Palomino,
and later as the Palomino horses. After Palomino's
death, some ten years after the conquest of Mexico,
these horses were turned loose. Later, when the ranchers
established themselves in Mexico and California, the
Palomino, due to his scarcity and beauty, was highly
prized by the wealthy ranchers.

This northeastern Nevada was truly a wonderful

stock country with its mountains and valleys, the mountains ranging in height from eight to eleven thousand feet, and the passes out of the valleys about six thousand feet. There was an abundance of small streams, and most of them ran through narrow valleys, with a strip of meadow on each side which had been sown to timothy and clover by the ranchers.

One of these valleys had a special appeal for me. Comprising about one hundred and forty acres, level as a table and with a slight elevation at one end, on either side of which was a spring, one almost boiling hot, which, by running it through a radiator, could have been used for heat during the winter months; the other ice cold. They came together at the edge of the meadow and ran through the center insuring open water for stock during the winter months. I had dreams of returning one day, filing a homestead on it, and building a ranch. The streams were filled with trout, and the surrounding country with game of all kinds, so one could live off the land indefinitely.

The first ranch I came to was a spread running about one thousand head of Herefords, or whitefaces as they were called. Men were just starting to put up hay, and were shorthanded, so I stayed on to help for about thirty days. I then rode on to Wells, which was a typical cow town with a liberal sprinkling of saloons and a red-light district where the ladies of the evening entertained the cowboys.

Wells was named by the emigrants because of the numerous springs scattered over the meadow northeast of town. It was one of the principal camping stops on the California Trail, and the members of many wagon trains camped here while their animals rested and put on fat and the travelers prepared for the long trip down the Humboldt.

Leaving Wells, we went down the Humboldt River to Elko, which was located in the midst of a prosperous stock country. After looking the town over, I arranged for a room and went to see Garcia about my riding equipment. As it was to be quite unusual, I had some difficulty getting my ideas across. I finally convinced him that I wanted an all-black outfit, using pigskin instead of the usual horse- or cowhide, as it was much lighter, more flexible, and certainly long-wearing, as I knew from my experience in England where most saddles were made of pigskin. The outfit was to consist of saddle, saddlebags, tapaderas, chaparreras, gun scabbard, and bridle, all to be as light as possible, consistent with strength, and to have a limited amount of silver ornamentation to accent the black. I had already ordered a silver-plated bit from England, together with a silver-plated chain curb as they were not available in this area. Specifications also called for the leather to be plain with no hand tooling, and, as he had to order special leather and make up the silver ornamentation, it was some time before the pieces of equipment were ready. However, they were surely worth waiting for, and, on Midnight they really stood out, as he made an excellent model. As for me, I ordered several black shirts and white silk mufflers, and these, together with a black broadbrim Spanish-type hat with chin strap and silver buckle, completed my outfit. My friend, Señor Valdez, would have liked to see me now, as I looked more like a Vaquero de Español than an American cowboy.

The following year was spent in Nevada, and my last job was establishing the grade for an irrigation ditch. There were no levels, transits, or engineers in the area, so our equipment was necessarily crude, consisting of a two-by-four, twenty feet long, with a sharpened leg at each end, also made from two-by-fours, and braced to the horizontal forming a truss for rigidity. On this we secured a carpenter's level, set to give us one quarter-inch fall every two feet. By starting from the point of delivery and making twenty-foot steps around the hills to the point of intersection with the creek, a satisfactory grade was established. By this time it was getting quite late in the fall, and snow had fallen on the higher elevations. I had been in Nevada over a year, and decided to return to Idaho by way of the Great Salt Lake Desert, and avoid the mountains. The first town, after leaving Nevada on my return, was Lucin, Utah, a shipping point on the Southern Pacific railroad. Upon arriving at Lucin, I found that one of the big sheep outfits using the desert for winter range was shipping its herds and wagons to the home ranch near Rexburg, Idaho, to feed them for the rest of the winter. They had everything loaded except one wagon and team, and did not want to go to the expense of another car for that. The foreman, with whom I had become acquainted, suggested that, since I was going to Rexburg anyway, I might drive them through and get paid for my trip. This appealed to me for many reasons: first,

MAP OF THE ROUTE—LUCIN, UTAH, TO REXBURG, IDAHO

PLATE III

getting paid for something I was going to do anyway, and, second, my camp would be right with me, and I would be assured of a good bed, plenty of food, and a good stove to cook it on. Then, too, it would be possible to carry sufficient grain for my horse as well as the team.

After we had agreed on wages and expenses, he wished me good luck, and in a short time was on his way with his sheep, wagons, and equipment. In talking to some of the old-timers as to conditions on the desert, I decided it might be necessary to make two or three dry camps, so, after procuring two water barrels in order that my water supply would be assured, and with my preparations complete, we started out in a northeasterly direction across the Great Salt Lake Desert, hoping to pass by the north end of the lake, and on to Stone on the Idaho border, a distance of about ninety miles. We arrived at Stone on the evening of the fourth day, and stayed over two days to give the horses a much-needed rest. From here on to Rexburg the towns were closer together, water was no problem, and I tried to arrange my camps so that we could travel about twenty-five miles each day. With this in mind, we went to Holbrook, then east across to Pleasant View. Here we turned north to Downey, then to Inkom, and, on the evening of the tenth day, we pulled into Fort Hall. Upon my arrival, I was, with the exception of the Indian agent, the only white person on the reservation. However, about an hour after I had made camp, a big prairie schooner from St. Joseph, Missouri, pulled in. Seeing all those Indians, and only one white man, the driver was quite nervous and asked if he could camp near me, which I was, of course, happy to have him do. He probably felt about like I did on seeing my first Indians, and pulled up so close that the canvases scraped together. I had ridden with several of these young bucks on horse roundup, so they knew me, and I felt quite at home. By this time they were all gathering around admiring Midnight, and, I believe, I could have traded him for half the ponies on the reservation. It is common knowledge, of course, that Indians generally have been mistreated by white men. However, these Indians, Shoshones and Bannocks, certainly were not when they were put on this reservation, as it is one of the finest areas for ranching and stock raising in the West, and many of them are very successful ranchers and cattlemen. Then, too, they had a fine Indian school, and all the younger ones, at least, were quite well educated for that period.

One of the first permanent settlements in Idaho, Fort Hall, was established as a trading post in 1834. The Hudson's Bay Company purchased the village, and later, in 1849, it became a military post. The original Fort Hall was on the banks of the Snake River, eleven miles west of the present site.

The Indians I had ridden with on horse roundup were exceptionally fine horsemen, and rode very well, but they had not yet learned to drive a team with any great success. The morning we were leaving, they had a six-horse hitch on a fairly large grain separator and were stalled in a small ditch. I don't believe they ever had more than one horse pulling at the same time; consequently they were not having much success. After watching them awhile, I suggested, rather facetiously, that they unhitch their team and we would pull their separator out for them. I was somewhat surprised and chagrined when they called my bluff. However, I had been called, and, as I looked at all those young bucks with grins spread all over their faces, I hoped I would not have to eat too much crow. I knew my team was good, as most sheep-camp teams were, as they had to get into and out of some pretty tough places, and we had been in some crossing the desert. With my fingers crossed, I hitched my team to the separator, and, when I called on them, they really did dig in and give it a good try. It took several efforts to get the load moving but move it they did, and in so doing saved me from a very embarrassing situation and changed those silly grins to awe and admiration for that old team.

After promising the man from St. Joseph, who was going to St. Anthony, that I would travel with him as far as Rexburg, and with good-byes and good wishes all around, we started for Firth where we intended to camp that night, and continue on to Rigby the next day. This left only ten miles to Rexburg, my destination. It had been thirteen days since we had left Lucin, and we had traveled two hundred and fifty-nine miles, some of it over rough and treacherous terrain.

My stepfather had come to the United States during 1905, and had settled in Rexburg. I had not seen him since leaving England, and wanted to hear about the family firsthand, so went to see him. He was preparing a home and told me he hoped to bring the rest of the family over in about two years. My older sister was staying in England, as she was getting married. After resting a few days, settling accounts with my employer, and visiting with my stepfather, I rode back to Idaho Falls to work the winter run at the sugar factory, which lasted three to four months. Upon completion of the run I returned by way of Blackfoot, and across the desert to American Falls and Rockland, the place I first went to when coming to this country.

I had met and become good friends with the family of Charles Howard, the man who had been kind to me and had made it possible for me to have a bed to sleep on my first night at American Falls. He was one of a large and quite remarkable family, and even today I am included in their family reunions, which they frequently have. They had a large ranch house and made me welcome for the rest of the winter, and, since they operated a small sawmill, I was able to help getting in the logs, which could only be done in the winter months. This family had a very lively sense of humor, so we had lots of fun. One incident I remember, which could have been serious but turned out to be funny, was the episode of the white-faced bull. Along the creek bottoms, where the cattle were fed, the willows grew very thick and were used for windbreaks. These willows were honeycombed with trails made by the cattle, and most of them had grown together at the top, making it impossible to get in on horseback. Somewhere in this maze was a renegade white-faced bull, and the owner had offered a reward of ten dollars to anyone getting him out so that he could be roped from a horse. This was a challenge to my friend Dewey and me, and then, too, ten dollars was a sizable amount in those days, so we decided to make the attempt. We started down one of the trails, cautiously, I must admit, as the growth was so thick there was no chance of getting off the trail, should we encounter the bull, and he charged as most ill-tempered renegades usually do. We had gone through several trails, with Dewey in the lead. He was the adventuresome type, and always looking for excitement. At any rate here we were searching these trails, which was like looking for a needle in a haystack, but at long last we came upon him in a little clearing. He was pawing the ground, and his eyes looked as big as saucers. As we expected, he dropped his head and charged. We did an about-face which put me in the lead, and, I suppose, we both had a wild idea that we might outrun him. This was soon discarded, and, when I happened to see a willow overhead that I thought might support my weight, I made a flying leap, grabbed one branch, and threw my feet around another. Dewey passed under me with the bull in hot pursuit, and the inevitable, of course, happened. A short way down the trail the bull overtook Dewey, struck him in the back, knocking him down, and charged on down the trail. I rushed over to where Dewey was lying, and pulled the prize faux pas of all time when I said, "Are you hurt?"

He looked up at me, and, with a pained look on his face, said, "Of course I'm hurt!"

I then compounded my boner by saying, "Did he step on you?" and again, with the same painful and disgusted look, he said, "What do you mean, step on me? The son of a gun danced on me for ten minutes!"

Fortunately, he was not badly hurt, as the bull's horns passed on each side of his body. It could have been much worse had be been gored. At any rate we got the bull out so that he could be roped, and we earned our ten dollars. We spent the rest of the winter logging for the sawmill, and doing various other things that had to be done.

CHAPTER VIII

The Life of a Sheepman

WITH THE ARRIVAL OF SPRING, WE RODE INTO TOWN and, while there, Carl Houtz, a sheepman who was looking for help, asked me to go to work. I informed him of my lack of knowledge regarding sheep, and that I was not anxious to become involved with cattlemen. After some discussion he convinced me that the trouble between the sheepmen and the cattlemen was over, as they were now allotted their respective sections on the leased government range. My knowledge of sheep was very limited but he had men who knew about that part of the operation, and my job was to see to the location of camps and to keep the men in supplies. He had two herds of about three thousand each, and was about ready to start the lambing season, which was to be a liberal education for me and one I shall never forget.

When I reported for work at the ranch two days later, one of the men there dropped everything and ran over to me as if he could not believe what he saw. As he came close, he said, "Habla usted español, señor?"

For a moment I was too surprised to speak. Then I remembered that I probably did look Spanish to him, and as I had picked up a little Spanish from Garcia and the Basques in Nevada, I replied, "Un pocito, si, no muy bien." He was obviously fresh from Spain, spoke no English, and needed to talk to someone.

This being the case, I said, "Buenos díaz, señor, cómo está usted?" to which he replied, "Muy bien, gracias, Y usted?" I replied, "Muy bien, gracias, cómo se llama usted?" He replied, "Me llamo Juan Martine Madarietta, servidor de usted," to which I replied, "Me llamo George E. Franklin, a sus ordenes."

He then started out in Spanish so fast he was completely over my head, and I had to break in with, "No entiendo, señor, hable usted mas des pacio, por favor." I liked this man instinctively, and knew we had to learn to communicate with each other, so I pointed to my horse and said, "Cómo se dice en español?" and he replied, "El caballo." Then I taught him to say, "How do you say in English?" and in this way we learned to use each other's language.

I soon observed that successful sheepmen always tried to keep the herd on the summer range at not more than fifteen hundred ewes and their lambs, in order to maintain them in good shape and allow the lambs to grow to good size. On the winter range the sheep were in the open, flat country, so three thousand could be handled quite satisfactorily. They spent about three months on the summer range, six on the winter range, and three at the home ranch feeding, lambing, docking, branding, and dipping.

In this part of the country, winter range is unlimited as there are hundreds of square miles of open desert country. Good summer range in the mountains is scarce, and limits the number of sheep that can be taken care of for the three summer months.

My employer's six thousand head were broken up into four herds of fifteen hundred in the summer, and two herds of three thousand on the winter range. He was a very successful operator, and had good tight sheds for the lambing season, instead of lambing on the open range as many sheepmen did. Consequently, his losses were much less than the average, which I am sure paid for the cost of the sheds.

On the winter range and feed ground, a sheep wagon was the home of the sheepherder and his camp tender for about nine months of the year. It was an ordinary wagon on the bottom, but projected out over the wheels to a width of about six feet six inches, and over this were wood bows covered with canvas. The entrance was from the front, but to one side, to allow for a small four-hole stove which was located in the left front corner of the wagon. Next to the stove, on the projection over the wheels, was a cupboard with the door hinged on the bottom in order that it might let down to form a table, which was supported by a leg, hinged to the outside of the cupboard door. In the corner back of the stove and upon the same shelf that supported the cupboard was the customary sourdough crock, without which the men would have no bread or biscuits. Across the back of the wagon was a full-size double bed with storage underneath. This storage was accessible either from inside the wagon or through the tail gate from the outside. Over the bed was a shelf for more storage, and, under this, a sliding window in the center of the back wall. On the summer range, packhorses and tents were used, and a heavy cast-iron Dutch oven replaced the little stove, as it was impossible to get the wagons to the higher elevations.

Since the bucks were put in the herd on the tenth of November, and taken out on the tenth of December, and, as the period of gestation is five months for sheep, lambs should now be starting to arrive. The sheep had to be held fairly close to camp, as lambs would be being born all day. As soon as a ewe lambed, she was moved to the shed area for protection against the weather. Sometimes a ewe would refuse to accept her lamb, so the ewe and lamb were marked with paint or chalk and the ewe was tied up so that she could not run off and leave the lamb. If, when she is turned loose later, she still leaves the lamb, we can catch and tie her, and find the lamb with the same number. Ewes identify their lambs by smell, and when a ewe, after smelling her lamb, bleats contentedly and lets him nurse, we know they are all right, and move them to the sheds. The first week is usually rough, because, when the bucks are first put in the herd, they are fat and the breeding instinct is strong, as a result of which lambs are born in larger numbers the first few days. After this, they calm down, and, toward the end of the thirty days, are much thinner, and inclined to be tired or lazy, so the birth rate dwindles as the days go by. Quite often a ewe would have a pair of twins, and some other ewe's lamb would die. The usual procedure was to catch the mother of the dead lamb and tie her by the front foot, then skin the dead lamb and stretch his hide over the

smaller of the twins, and set him down beside the tied
ewe. If left alone, she would, after a while, smell the
hide of her own dead lamb, and claim the twin as her
own. Next day she would be untied but hobbled so that
she could not run faster than the adopted lamb. When
the lamb was strong enough to keep up with her, the
hobbles were removed, and she usually decided to keep
him. As the lamb grew he finally broke out of the
dead lamb's hide and it fell off, but, by this time, the
ewe would know her adopted lamb by the sound of his
voice.

About the end of the month there were very few
ewes left in the drop band, and, when we had no lambs
for three days, it was time to think about docking. In
the corner of the large corral we built a pen about ten
feet square with a two-by-ten plank nailed to the top
rail. The ewes and their lambs were then driven into
the large corral, and the men walked among them, pick-
ing up the lambs and dropping them into the small pen,
and when it was full, they started docking. One herder
stood in the pen, and another stood outside the fence
with a sharp pocketknife, made especially for this purpose and called a "stockman's special." It consisted of
a skinning blade, a castrating blade, and a general-pur-
pose blade. Another man stood beside him holding a
sharp pair of sheep shears. Nearby was a pail of melted
lard and turpentine with an applicator, consisting of a
stick with a piece of sheepskin tacked to one end. The
herder in the pen would pick up a lamb, grasp the front
and rear legs of one side in his left hand, and of the other
side in his right. He would then rest the lamb on its

back across the two-by-ten plank on top of the fence, with his tail toward the man outside. If it was a male, the herder would castrate it.

The man with the sheep shears would then cut the tail off about an inch from the body, and apply some of the melted lard and turpentine to the stump, slowing the bleeding and helping the wound to heal, as well as keeping the flies away.

The lamb would then be turned around with his head facing the outside, and he would be earmarked in accordance with the registered brand. After the last lamb had been docked, the ewes were turned loose to find their own. They bleated long and loud and the lambs replied in their high-pitched voices, creating a din that could be heard for miles.

There were still shearing, dipping, and branding to be done before we started to trail for the summer range. The shearing was usually done by a special crew brought in for that purpose. They started in the south and worked north. The dipping was usually accomplished by the use of an elongated tank or flume about two feet wide and one hundred feet long, the sheep being forced to go through a chute into one end of the tank, then they had to swim the hundred feet. This swim through chemically treated water destroyed all the ticks and other vermin common to sheep.

Branding was done with a wooden brand and paint, different brands and colors being used by different owners. With all these things out of the way, we were now ready to start for the summer range which was about seventy-five miles distant.

My employer, Carl Houtz, was very good to work for, as he believed that if he took good care of his men they, in turn, would take care of him. He was one of the few who always maintained two men in each camp, thinking that, if one became ill, he could lose a lot of sheep, which would more than pay for the extra man. Later, two men in a camp became mandatory. I had really tried to work for his interest while in his employ, and even though there were other men older and more experienced than I, insofar as sheep were concerned, he asked me if I thought I could handle the outfit on the trail and summer range. I pointed out these obvious facts, and he replied, "It depends upon what experience you need. True, I have men who know sheep, but, aside from that, they are lost, and I want someone that can be depended upon to see that the sheep are moved around as they should be to get the best grazing." He had to stay and run the ranch, but told me he would write to stores along the way to give me what supplies were needed. When I asked about credentials, he said, "Your horse Midnight is all the credentials you need, as you are both known in three states." It was true that we had become well known, had been entered in many races and parades, and had covered quite a lot of territory. Midnight had, through the past year, become quite gentle, possibly because I had always managed to have some good oats, lump sugar, or some other delicacy for him, so it was not necessary to hobble him any more. All I had to do was whistle and he was there ready, willing, and able.

We decided to start one herd of three thousand about

three days ahead of the other in order to have time to separate them into bands of fifteen hundred each for the summer range. This was to be done at a ranch the boss had homesteaded before it had become a reserve. It was ideal for this purpose, as it was right in the center of the range he had leased from the government. There were good corrals and chutes, so separating the sheep would be no problem.

With all the plans taken care of, we started the first herd on the trail, planning to go around the south end of the Fort Hall Indian Reservation, and on past Mc-Cammon to the vicinity of Lava Hot Springs, where we would turn north past Bancroft, and on into the Pebble Creek Basin where the ranch was located. From there we would distribute the herds around the hills wherever grazing and water were best. Most of the sheepmen in the area had Merinos, but Mr. Houtz, had been using Cotswold bucks on part of his herd to build up a bigger and better type that would bring more on the market.

These sheep were much bigger and longer legged than the Merinos, and traveled much faster. Consequently, instead of having trouble with the drags as we usually did, we had trouble with the leaders and were constantly throwing them back. These sheep knew where they were going, since they had been there before, and knew that good grass and water awaited them.

Since about twelve hundred of the three thousand were Cotswolds and the rest Merinos, it did present a problem, and, if we were not careful, the herd would be spread out for ten miles, which could not be tolerated.

We were working the dogs very hard; their feet were becoming sore, and, in order to keep them going, we had to make little leather boots for them. In desperation I rode to the head of the herd and kicked dust in the faces of the leaders with a .30-30 Winchester I always carried in my saddle scabbard. This helped, as I could ride off to one side, and, when the leaders started going too fast, a well-placed shot would turn them back.

About our third day on the trail, a problem presented itself that could have had serious repercussions. Fortunately, by this time, both the cattlemen and sheepmen had calmed down a lot, due to the practice of allotting different sections of range to each, and they would, if hungry enough, stop at each other's camp to eat. On this particular occasion our camp tender had gone on ahead with the wagon, as he usually did, to make camp for the night. When I pulled in, with the herd not too far behind, I noticed that he was cooking beef for supper, and, at the same time, I saw two cowboys coming over a ridge not too far distant. I knew they were going to stop and eat, as it was a long way from any ranch, and near suppertime. I was also afraid he was cooking their beef, but to make sure, I said, "Where in hell did you get that beef?" Being inside the wagon he had not seen the cowboys and said, "Well, we have been eating mutton for months, and I just had to have a little baby beef, so I got myself a yearling. He is dressed, and in the back of the wagon."

By this time the cowboys were very near, and I had visions of what might happen if they caught us eating their beef, so I said, "Keep your mouth shut, and let

me handle this." I grabbed a pepper can, worked the top
loose, and, when they rode up close to the door, pushed
the lid off with my thumb, spilling the contents of the
can all over the meat. Apologizing for my clumsiness,
I turned it back to the camp tender, with instructions
to dump it in a pail so that it could be washed off for
the dogs, then cook some more lamb chops for the boys,
as they looked hungry. The crisis over, I visited with
them until supper was ready, when they ate and continued on their way.

Arriving at the ranch on the summer range, we separated the herds into four groups of fifteen hundred
each, and located them in their respective camps. This
summer range in Pebble Creek Basin was one of the
most beautiful places I had ever seen. Completely surrounded with mountains, it was almost like a huge basin
as its name implied. Pebble Creek ran through the
center and out through the canyon where we had entered. The water was ice cold from the snowbanks, and
crystal clear, running over a solid bed of pebbles. There
was an abundance of mountain trout for the taking,
and the surrounding country was a hunter's paradise,
containing game of all kinds from grouse to black bear.
I pitched our tent at the edge of a little clump of pines,
which effectively broke the wind and gave us a good
deal of protection.

After we were established, and in the days ahead, the
sheep started to move out to graze early in the morning,
shaded up during the heat of the day, going out again
toward evening. There was very little to do, so we made

up for the strenuous time we had on the lambing ground and coming up the trail.

During this time I had become better acquainted with my friend Juan, and we had decided that I would stay at his camp and act as camp tender. This meant more work for me, as I also had the overall supervision of the other camps, as well as going to town every other week for supplies. However, Juan said he could do most of the camp work, and he did want to learn English, as I wanted to learn Spanish. The second day I decided to ride to the other camps to see that everything was in order, and when I returned in the evening Juan had shot a nice bunch of grouse which he had prepared for the evening meal. If you have never eaten food prepared in a Dutch oven, you cannot appreciate its value as a camp utensil. A Dutch oven is a heavy cast-iron vessel or pot with a lid having a turned-up flange to hold hot coals. The food to be cooked is placed in the oven, covered with the lid, and placed in the fire with hot coals all around and over it. This keeps all the flavor in and results in food that would be the delight of any gourmet. In preparing the grouse, Juan had rendered a few strips of bacon to grease the oven, so that the birds would not stick. He had then covered the bottom of the oven with breasts of grouse, over which he had laid strips of bacon solid, and put it in the fire to cook. When cooked, he had poured over them a can of small French peas. This, together with mashed potatoes, sourdough biscuits, butter and honey, as well as good coffee, made a very enjoyable meal. As we sat under the trees, with the pine smoke from our campfire

tickling our nostrils, we listened to the rippling of
Pebble Creek a few yards away. It was a most wonderful evening, and we enjoyed it to the full.

Now that our camps were established, life was much
easier, and it seemed like a vacation to me. Juan enjoyed
doing most of the camp work and cooking, which left
me a lot of time to think, and I found myself looking
over my life in retrospect. I had accomplished most of
the things I had set out to do, but I now realized that
the glamour was wearing thin, and the desire to be with
horses, cattle, and sheep was not as strong as it had been.
I still loved my personal horse, Midnight, and looked
upon him as a true and trusted friend, but I could see
men getting old at thirty as a result of the hard life
they led, and I decided it was not for me. Then, too, I
wanted to improve my education, and since, even as a
child, I had been mechanically inclined, I decided to
take an International Correspondence Course in mechanical engineering and use the spare time I now had.
I also managed to get some good books, and a diction-
ary which enabled me to improve my vocabulary.

The summer passed all too quickly, and, in order not
to get caught in the high country by snow, we started
for the winter range, which was across the Snake River
from American Falls. This time we trailed across the
Fort Hall Reservation, as we were returning in a different direction, and it saved many miles. However, we
had to get permission and an Indian escort, as we were
not permitted to camp on the reservation with stock
overnight, making it necessary to camp on the east side

one night, trail across, and camp on the west side the next night.

We crossed the Snake River at American Falls and continued north to our winter range which was ideal for this purpose. There was, of course, always the problem of water and I remember one day, as I was looking for some shallow depressions that sometimes held rainwater not yet evaporated, being rudely awakened from my daydreams by a sizable bunch of ducks rising out of the sagebrush. At first I thought it must be a mirage, but, when a second bunch rose up, I knew they were real and that where there were ducks there had to be water. I had ridden this desert many times and knew of no water closer than twenty miles, so I was, quite naturally, astonished to find ducks here. After I rode a little farther, the ground looked damp and then, through the brush, I could see water. I did not know from where it came, nor did I care. Water was what we needed and here it was.

Next morning I rode around the lake, which was quite large, to determine from where it had come. After some investigation, I found that a huge irrigation canal, which started from the Snake River near Blackfoot and roughly paralleled the river for many miles across the desert, had broken and flowed into this depression, causing the lake. We moved our wagons in, one on each side of the lake, and were assured of water for as long as we would need it, or until snow fell, when the sheep could use it instead of water, and as long as the snow was not too deep they could graze until such time as we trailed them to the ranch to feed. For ourselves and horses, we

AN EVERYDAY PROCEDURE IN CATTLE COUNTRY

PLATE IV

MIDNIGHT IN THE WILD BUNCH

Midnight as outfitted by Garcia

MIDNIGHT AS OUTFITTED BY GARCIA

PLATE VI

GEORGE E. FRANKLIN IN WORLD WAR I

As the engineer certifying planes for flight, he accompanied the
pilot and rode in the observer's seat. Qualified personnel were in
such short supply at the time that only one plane in four was flight-
tested before being shipped overseas.

PLATE VII

FRANKLIN AIRCRAFT ENGINE

PLATE VIII

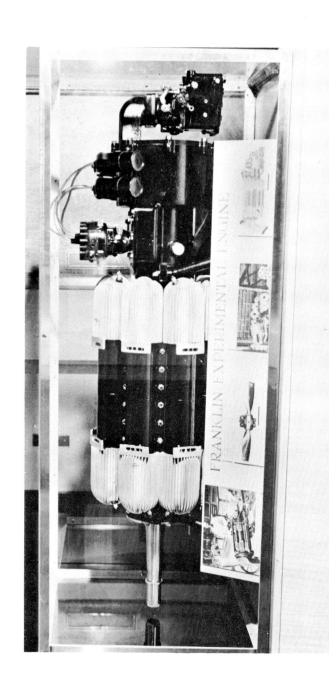

FRANKLIN EXPERIMENTAL ENGINE

Engine exhibited as part of the Aviation History Project, Montana Historical Society, Helena, Montana. The engine is of the barrel type and has sixteen cylinders mounted in opposed pairs. It operates on the two-cycle principle and is air-cooled. Fuel is supplied by a supercharger which is so designed as to give sea-level conditions at fifteen thousand feet altitude.

FRANKLIN EXPERIMENTAL ENGINE

PLATE X

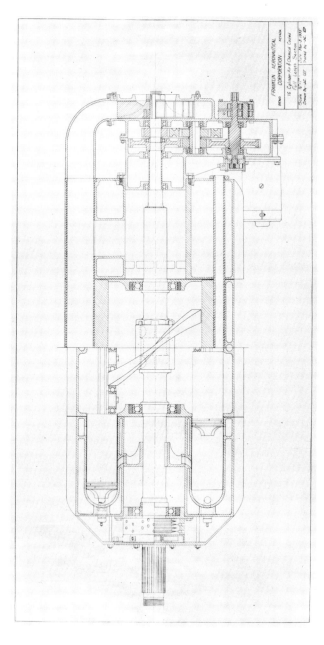

FULL-LENGTH ENGINE SECTION

Drawn by G. E. Franklin, James Crawford, and Edward Parmenter

PLATE XI

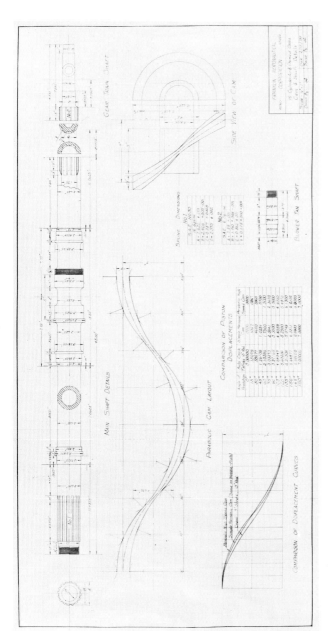

CAM LAYOUT
Drawn by James Crawford
PLATE XII

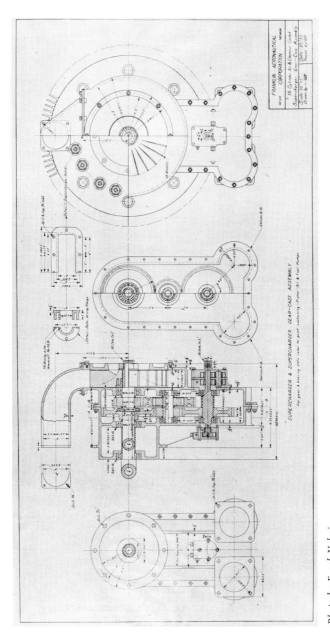

SUPERCHARGER GEAR-CASE ASSEMBLY

Drawn by G. E. Franklin

PLATE XIII

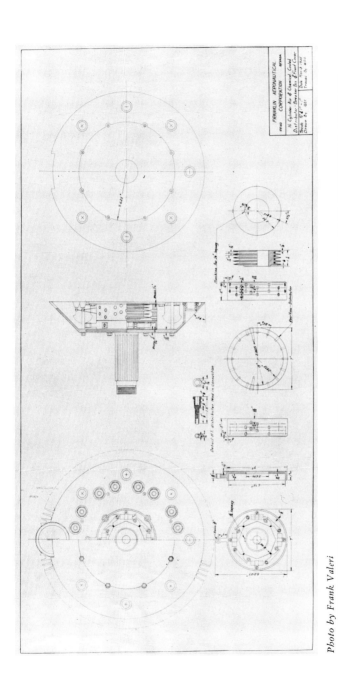

DISTRIBUTOR-BREAKER BOX AND FRONT COVER FOR AIR AND CHEMICAL COOLED
ENGINE.

Drawn by G. E. Franklin

PLATE XIV

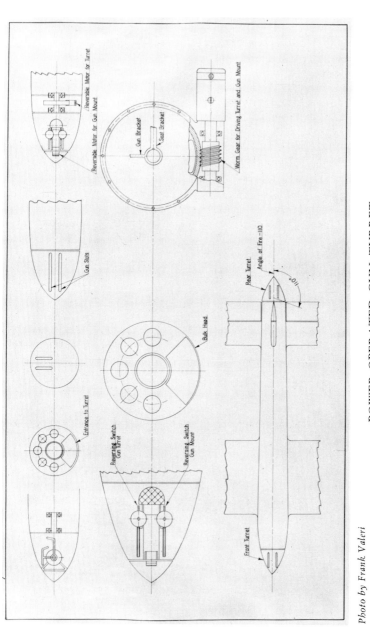

Reversible Motor for Turret

Reversible Motor for Gun Mount

Gun Bracket

Seat Bracket

Worm Gear for Driving Turret and Gun Mount

Gun Slots

Bulk Head

Entrance to Turret

Reversing Switch Gun Turret

Reversing Switch Gun Mount

Rear Turret

Angle of Fire =110°

110°

Front Turret

Photo by Frank Valeri

POWER-OPERATED GUN TURRET

Submitted to the Air Corps and to the N.A.C.A. by Congressman James G. Scrugham, representative from Nevada. The sketches are intended only to show the application of the device. No attempt has been made to work out any engineering details as these would have to be applicable to the particular plane on which the turret was to be installed. With this installation it is possible to rotate the turret through 360 degrees and the gun mount through an arc of 110 degrees, reversing the direction of rotation at will. Drawings by G. F. Franklin.

propped up the wagon tongue and suspended a wash-tub from it, in which snow was melted for our use.

This desert was unique in that it had several large lava beds upon which nothing grew except some cedars which had taken root in the crevices that had blown full of sand. Other than this, it was just black lava. These enormous lava beds were almost impossible to penetrate, except on foot, and, even then, the going was extremely rough. Some of the crevices were so deep one could drop a pebble, and wait a considerable time before hearing it hit bottom. The desert between the lava beds was, however, very good winter grazing land, as the grass and vegetation grew quite tall during the spring and early summer, and was cured as it stood, making fine winter feed. Another factor making this area desirable for the sheepmen was the railroad out of American Falls, going west. The grade was quite severe, causing the freight trains to slow down considerably, so that the men could board them and throw off coal, which was picked up later and hauled to camp, where it was greatly appreciated after burning sagebrush.

Since my friend Juan and I had spent almost a year together, we had grown to know each other quite well and enjoyed our companionship as we more or less liked the same things. He had, by this time, learned quite a little English and I had improved my Spanish, so that we conversed quite well.

One day, while he was reminiscing, he told me about his girl in Spain, and how his parents had another girl picked out for him, which was the custom at that time, so he left home and came to the United States.

Juan was a handsome man about six foot one inch tall, blue-eyed and fair-complexioned, and having a great physique. I would have thought he could have taken his pick of the girls, so I was not in the least excited when he showed me a photograph of a very plain girl, and said, "This is my Juliana." Before I realized it, I said, "Why, Juan, she is homely!"

His immediate response was, "Ah, you don't see her with my eyes!"

We had formed a habit of Juan speaking as much as possible in English, and I in Spanish, so I replied, "Sí, es verdad, no veo con su ojos, pero ella no es bonita." ("Yes, that's true, I don't see her with your eyes, but she is not pretty"). Not too long after this he received a letter telling him that Juliana was getting married. He took it quite well and later that winter met a girl in Jordan Valley, a Basque community near Boise, Idaho, where he had gone on vacation. She was, I thought, more suitable for him, tall, blue-eyed, and fair-complexioned as he was, and they made a very striking couple. We both worked two more years. Then Juan left for Jordan Valley to start in the sheep business for himself, and to marry and settle down.

During the past two years my work had been comparatively easy except for lambing and shearing, and I had been able to spend considerable time on my engineering studies, which were now at about the halfway mark, as it should be. It had not been easy, with no teacher to instruct or advise me, and I had to dig it out the hard way.

The sheep and all the equipment were now at the

ranch, and there was nothing to do but feed, and ample help for that, so I decided to go to Idaho Falls and the sugar factory. Through the years this had been my program for several reasons: first, it paid well when work was scarce on the ranches; second, it was a good-sized town where a good show could be seen; and third, there was a Y.M.C.A. with a well-equipped gymnasium where one could get in good physical condition.

CHAPTER IX

Jarbidge Episode

THE WINTER WAS GOING ABOUT AS USUAL WHEN, DURing December, there was a lot of publicity about a rich gold strike at Jarbidge, Nevada. My roommate, Lon Forsee, was from Globe, Arizona, and had had some experience in mining, so, after discussing it at some length, we decided to try our luck.

We were more or less conservative and, unlike most young men of those days, had a little money in the bank, so we felt it would be an experience even if we did not strike anything. We decided to outfit at Twin Falls, which was the last town of any size on the way in. From Twin Falls we joined a freight outfit and traveled by way of Hollister, Rogerson, Three Creek, and on to Kitty's Hot Springs, commonly known as Kitty's Hot Hole, a distance of about ninety miles. This was at the rimrock and was the end of the wagon road. In fact, it was the end of the trail, as there was nothing beyond this point, and supplies were being piled up, and covered with tarpaulins, waiting for a pack trail to be blasted down the rimrock cliff so that they could continue on to Jarbidge. Not knowing how long this would take, and anxious to get to our destination, we stored the bulk of our supplies with the packtrain company until such

time as they could get through, made arrangements
with Kitty's Ranch to board our horses, and, with each
carrying a pack of about sixty pounds, started for Jarbidge, about eighteen miles distant.

We had started at daybreak and did well to get there
before dark, as the going was rough and we had no
snowshoes. There were a few people ahead of us who had
come in from the Nevada side by way of Charleston and
Deeth, though, even on this side there was no wagon
road between Jarbidge and Charleston, and the only way
was by horseback and packtrain. We were exhausted
when we arrived, and would have had a very rough time
had it not been for the kindness of one Bob Byrne, a
big friendly Irishman, who had preceded us, and had a
camp established. As was the custom in those days, he
had a five-gallon kerosene can with a wood frame around
the top, and a round bar across to lift it with. In this
he had cooked a regular Irish stew with about everything in it—beef, potatoes, carrots, onions, tomatoes,
and so on. He served both of us a big bowl and it certainly did a lot for our waning strength and courage.

We had packed a small tent to protect us from the
weather, some blankets, and a limited supply of food
to last until the packtrain came through with the rest
of our supplies. Bob, seeing our limited outfit and supplies, and, I suppose, realizing that we were neophytes
at this business, insisted that we stay with him until our
supplies arrived. He had ample room, good equipment,
and, best of all, a good Sibley stove to keep us warm.
This was all possible because he was able to bring it in
on packhorses. I shall never forget that night, January

6, 1909, just six years to the day after I had landed at
Boston.

Before leaving Twin Falls, we had, fortunately, become acquainted with a man who had been with Bourne
when he made the original strike, and who had located
a claim on Bear Creek, just off the main stream of the
Jarbidge River. He gave us permission to camp on his
claim, as he felt it might give him some protection if
someone was living on it. So, while waiting for our
supplies, we started to level a spot under a big cotton-
wood tree that we thought might give us some shade
during the summer months. It was about ten days before the packtrains started to arrive, and we could set
up housekeeping for ourselves. We soon discovered
that my partner was not much of a cook. When he attempted to cook a pot of rice, he had every utensil in
camp filled before he got through, so we decided then
and there to divide the work, he would carry the wood
and water, and wash the dishes, while I did the cooking.

With the opening of the pack trail, things began to
hum, and the ingenuity of the packers was something
to behold. They even brought in pianos and hotel-size
ranges on muleback, as well as lumber and many other
things I would have thought impossible. With the large
things such as pianos, and ranges, they would pick two
husky mules, lash a pole on each side of the packsaddles,
and load the piano or range between them. Lumber
they would lash to each side of the animal, and let the
other end drag, and as a result about twelve to eighteen
inches would have to be cut off to square the end before it could be used.

More people were arriving every day, so we soon had a sizable town along the Jarbidge River, and on Bear Creek.

By this time most of the men were frantically scouring the hills and locating mining claims which was ridiculous as there was as much as thirty feet of snow in places, and, during the autumn, when we went over the same territory, their location notices were thirty to thirty-five feet in the treetops, and nothing in the way of mineral to locate anyway. But that was typical in a gold stampede.

This tramping around the mountains during the winter was also very dangerous as the peaks rise to an elevation of eleven thousand feet and comprise four or five extinct volcanoes. These had eroded on the northeast side, from which flowed small creeks. Quite often, winds of high velocity swirled around these peaks, causing the snow to comb out sometimes as much as twenty feet from the vertical wall of the crater, and, unless one knew the country, this could be very hazardous. During February, two men were prospecting in this high country, one wearing skis, the other small webs made by the Indians, and called bear-paws. Suddenly the one wearing skis realized that his partner was not with him, and, upon retracing his trail, found the man had walked too close to the edge of the combing, which had collapsed, dropping him to the bottom of the crater twelve hundred feet below. There was a one-thousand-dollar reward placed on him, but his body was not recovered until the following June, when he was taken out of the snow and ice in perfect condition, as he had been frozen solid. This accident caused many prospectors to think twice before venturing into this area, at least until the snow had melted.

My partner and I had decided when we first arrived that we would not attempt to prospect until later, and, as there was plenty of work at good wages, we went to work helping to build the town. Jarbidge (Ja-habich the Indians called it, meaning the Devil, or an evil place in which they would not camp overnight), was located mainly on the Jarbidge River, and was a one-street town, except for a short street branching off on Bear Creek. It was on the bottom of a deep canyon where, even in midsummer, the sun never penetrated until about nine o'clock in the morning. The place was surrounded by mountains, some of which reached an elevation of eleven thousand feet. During the winter months they were heavily blanketed with snow. In the summer it was a thing of beauty, with the crystal-clear river full of trout, and the surrounding country full of game of all kinds, including antelope, deer, cougar, and wildcat, as well as many smaller animals. It was also a forest reserve, and thousands of sheep were grazed in the surrounding hills during the summer months.

With the coming of spring, and the improvement in transportation, although we had no wagon road as yet, more of the usual riffraff of a new mining camp started to arrive. The prostitutes and their pimps were the first to come, followed by the card sharks, claim jumpers, gunslingers (of whom, fortunately, there were few), and human jackals, all preying on the legitimate people of the town. Neither my partner nor I came in contact with this element to any great extent as we stayed at our camp evenings, reading and studying, but we could not avoid brushing elbows occasionally, and, for this reason, never carried our guns downtown, which was true of most law-abiding men, because it was just an invitation to the gunslinger type to start an argument. Then, too, there was a more or less unwritten law of the West that frowned upon drawing on an unarmed man. Consequently, if he insisted on quarreling, the gunslinger almost had to take off his guns and meet you on even terms. This most men, and particularly miners, did not mind as they were a hardy lot and could take care of themselves. Finding myself in this environment, I remembered the advice my mother gave me, as I was leaving for the States. Remember, she said, "There is only one person in the world you cannot get away from, and that is yourself, so whatever you do you have to live with yourself. Second, do not ever look for trouble, and, if it comes to you, try and talk your way out of it. If this is impossible, keep yourself in shape to finish whatever you begin."

I was now fully grown and weighed one hundred and seventy pounds, none of which was fat. Every winter considerable time had been spent in the gymnasium, boxing, wrestling, and working with the bars, rings, and ladders. In addition to this, a dentist friend from Pocatello, Dr. Sherbourne, who was an amateur wrestler, had spent a month in our camp each summer we were in Pebble Basin. He was very good and taught me many things including the art of jujitsu, so I felt that I had followed my mother's advice even to avoiding trouble, but that I could handle it if it came.

This boom-camp atmosphere was new and exciting to me, something new every day. One day a prospector came in and reported that he had found the skeleton of a man, lying at a spring, just as if he had been drinking and died there. Later his watch was found where he had hung it on a small tree. Through the years the tree had grown around it until there was only a little of the metal exposed, so this became known as Dead Man's Spring.

Not long after we had established our camp, a man came in from the Nevada side, built a tripod, and covered it with canvas, next to us, and a tent for himself next to that. He came over and introduced himself, and I believe his name was Van Pelt, or some name very similar. He must have known about us because he handed us a letter from the owner giving him permission to use the space he occupied. About that time there was a group of promoters trying to sell lots they did not own, and, unfortunately, they picked on Bear Creek, which was privately owned and few people on it. These people one evening succeeded in getting our friend Bobby Byrne, very drunk, and, it seems, he turned from a big friendly Irishman to a fighting Irishman. Taking advantage of his condition, these scoundrels succeeded in getting him to tear down and remove the tripod next to our tent. It was a beautiful moonlight night about eleven o'clock, when I heard him coming down Bear Creek singing and swearing. I thought nothing of it until he stopped and removed the tripod, carrying it with him down the street. The man next door called to him saying, "Better bring that back, I may need it." Bobby dropped behind a big rhyolite boulder, and, swearing a blue streak, opened up with a German Luger, the bullets ripping through our tent just above our heads. After a brief interval he apparently raised up to fire again, and when he did, got a .38-55 Winchester bullet right between the eyes. My partner and I were very upset about this, as he had been so good to us on our first night in camp, so I made up my mind, then and there, that I would never allow myself to become intoxicated. At the trial, two days later, we learned that our neighbor was sent to camp as a Deputy U.S. Marshal.

It was now late spring and we thought we would try our luck at a little prospecting. So we started up the Jarbidge River toward the divide between the headwaters of the Jarbidge and the Marys River, which flowed south to Deeth, Nevada. Before leaving town, we had heard of the Jones Brothers who ran sheep on the reservation, and apparently realized that the mining would destroy their summer range, as mining took precedence over stock. At any rate we had heard that they had ridden all over the country, staking mining claims, which would hold them for ninety days, and insure the range for that summer. Later it was brought to our attention in a somewhat humorous vein when we came across a location monument some wag had built beside the trail. He had done a beautiful job, and had met all requirements—rock base of the proper proportions, a post at the center, hewn with meticulous care to the

proper dimensions, and a tobacco can for the location
notice, which was also proper. As a final touch, he had
described his claim as follows, "I claim ten feet east and
ten feet west and Jones and Jones, they claim the rest."

We continued on upriver past Snowslide Gulch,
which was completely denuded of all timber. Years before a bad slide had deposited a part of the mountain-
side in the bottom of the canyon, and the annual slides
since had allowed no trees to grow. Later that year my
partner and I had quite an experience in Snowslide
Gulch. We were returning from the high country. All
the ridges were bare but we decided to take a shortcut
across the gulch as the snow was packed hard and snowshoes were not required. We did not expect any trouble, but when we were a little way out on the snow, we
heard a loud report, and across the canyon saw a huge
crack opening up. It was following around the head
of the canyon and we had to get back to high ground
in a hurry or be caught in the slide. Fortunately we
were both in good physical condition, and I know we
were both praying as we raced that widening crevasse.
The race was very close when a crack opened behind us
as we reached higher ground, and, as we watched that
whole canyon of snow in motion, a chill ran up my back
as I realized that we had come very near being in the
middle of it.

Our camp that night was just below the divide on
what appeared to be an old Indian campground. Ar-
rowheads, and the materials from which they were
made, were all over the place. There was a small circular lake and some pine trees with plenty of firewood,

so we had a comfortable camp. Next morning we went up on the divide from where we could see for miles down Marys River, which was a beautiful sight. For the next few days we prospected on both sides of the divide, and, finding nothing, decided to go north along the ridge toward the Matterhorn which has an elevation of 10,889 feet.

Several times from this point, when conditions were right, I have been able to see the Grand Tetons in Wyoming approximately 260 miles away. From this same point, by turning northwest, one can see the canyons of the Jarbidge, and the west fork of the Bruneau, to the confluence of the Bruneau and Snake rivers, which graphically portrays what happened when they were created. It is simply a huge crack or crevasse, and, whenever there is a projection on one side, there is a corresponding indentation on the other, and, even after all these years of erosion, it is still visible and indicates what a tremendous upheaval it must have been at the time it happened.

Along the ridge near the Matterhorn we found some outcroppings that looked interesting, so we decided to camp and prospect the area. Just below timberline we found a little lake and some scrub pine, where we set up our teepee tent we had brought along for occasions of this kind.

We prospected until late fall, and the probable arrival of the first snow, when, not wishing to get caught up there in bad weather, we returned to our permanent camp.

After resting a few days we were successful bidders

on a contract to run a tunnel one hundred feet into the hill above town. This would enable us to work underground, out of the weather, for the rest of the winter, and still live in our town camp. We did have a very steep climb every morning to get to our work, but in the evening, when we returned to camp we used a very light, homemade toboggan we had strapped to our backs going up, and came down that mountain at about sixty miles an hour. I had to depend on my partner's judgment, as I had never worked underground and did not know a single jack from a double jack, but I soon found out that a single jack was a four-pound hammer used with one hand, while a double jack is an eight-pound hammer used with both hands.

Starting our tunnel was somewhat difficult, and we had to run an open cut for about fifteen feet before getting underground. However, we managed it before snowfall, and, as this was my first experience underground, I had a lot to learn, as the successful breaking of ground depended upon the placing of holes, order of firing, and many other factors it was necessary to know about. This was all handwork, as the camp had not yet developed to the point where compressors and machinery had been brought in. In due time we cut the quartz vein that was indicated on the surface, and I learned about salting mines. I had heard considerable discussion as to the various methods of salting a mine, but this was my first actual contact with it. This man was clever, and we only became suspicious when, instead of the powder we usually received in boxes, he brought a bundle tied with cord and told us to use this powder in the next round of shots, explaining that the store was temporarily out of stock. This we found hard to believe, and, when the opportunity presented itself, we examined one of the sticks and found that he had used a miner's candlestick, which is usually made from a quarter-inch steel bar, sharpened to a point on one end and formed into a loop on the other, and, of course, a socket for the candle. This device had been used to make a hole down through the center of the sticks of dynamite. These holes had then been loaded with high-grade ore from other proven mines, and, when the explosion occurred, it was blasted into the surrounding quartz, and would appear to anyone but an expert to be in place. On the basis of these specimens many people were buying stock in a mine that might or might not develop into something worthwhile. Naturally we were quite perturbed, but we were just working in the mine, and there was nothing we could do to change the situation.

Whether or not this particular mine ever developed in later years I do not know. We did, while working there, have a somewhat legendary and famous visitor in the person of Death Valley Scotty.

Since no post office had as yet been established, we had all been paying an individual twenty-five cents per letter to bring them in for us. Now, however, a contract had just been awarded to carry mail from Deeth, which was a tough route, as it had to come in over the Coon Creek Summit, where the snow was usually very deep during the winter months. There was no wagon road, so it had to be carried by packtrain, and here, too, the carrier, one Bill McLaughlin, showed his ingenuity.

He could never have made it through the deep snow
using ordinary methods, so he broke a number of small
ponies to wear snowshoes, and, strange as it may seem,
they became very good. It made pacers out of them,
since they had to advance both feet on one side, and
then both feet on the other in order that they would
not step on their snowshoes, which were necessarily of
the short bear-paw type. Then, too, they had to spread
their feet apart for the same reason. At any rate, we
got our mail via Bill McLaughlin's snowshoe-equipped
packtrain.

The wagon road from the Idaho side was now nearing
completion, so that heavier equipment could be hauled
in, and there were rumors of several mills being built,
one at the Bourne, the original strike, the Pavlak, Blus-
ter, and others. Since there was no power into the area,
steam boilers and steam-driven dynamos had to be
used, and, of course, they were wood burners, as there
was no coal available. This being the case there would
be a demand for cordwood for the boilers, and timbers
for the mills, so we gave up our mining operations temporarily and concentrated on locating some good timber, which we found on top of the ridge above the
Pavlak mine. This area comprised both good standing
timber suitable for mill framing and fire-killed trees
that would make excellent cordwood for the boilers.
We had to take in another partner, as I was not yet of
age and could not file on a claim, and, according to
Nevada law, it required at least two men to get the
claims located in one compact group.

As soon as our claims were staked, we built a good

tight cabin located at the edge of the timber in a small meadow. At the rear was a lean-to for our horses, as they had to have protection during the winter months. Our new partner, Gordon Bettles, who was related to old Bettles of Klondike fame, was an expert woods-man, which fitted into our plans perfectly. He could fell a tree anywhere he wanted, even against the natural lean of the tree. One day the Prunty Boys came through from Charleston, where they had a mine and a large cat-tle ranch. They were the typical hillbilly type, and, I believe, were originally from Virginia. One of the brothers also claimed to be an expert woodsman and quite a discussion took place as to who was the better man. Finally Prunty bet Bettles a five-dollar hat that he could not fell a tree and drive a stake where he, Prunty, set it. The bet was taken, and Prunty, after surveying the tree from all angles, set the stake in what he thought was the most difficult position, away from the natural lean of the tree. Bettles looked the situation over very carefully, and then with saw and ax cut out a wedge facing the direction he wanted the tree to fall. He then went to the back of the tree, started his saw cut, and with the aid of wedges forced the tree over to fall directly on the stake driven by Prunty, thereby win-ning the bet and the hat.

A road was now being built from Charleston to Jar-bidge and had almost reached the summit near our cabin when the Prunty Boys came through. We arranged to trade them a set of house logs for beef and honey, to be delivered when the road was available. The road from Charleston to Jarbidge is truly beautiful, as it passes through some old placer diggings, gently ascends the mountains to Coon Creek Summit, where the road remains at an altitude of over eight thousand feet until it reaches Bear Creek Summit at 8,488 feet. From this summit to the Jarbidge River is a descent of over two thousand feet in five miles, and when one reaches the river, it is less than two miles to Jarbidge, in an alpine setting at 6,200 feet.

As we expected to take advantage of the snow to slide the mill timbers off the mountain to the millsite, we started to get them out first. The trees were felled, and then hewn to the desired dimensions with broad axes and adzes, and piled up on the edge of the runway to await the coming of the snow. The cordwood could be taken care of when required. We all worked very hard to complete our contract, which was quite satisfactory, and then returned to our town camp for a much-needed rest. Some time before this we had heard of a rich strike in the Yaqui River country of Mexico, and had decided to investigate. At that time the Yaqui Indians were restless, so we had to be prepared, and six of us standardized on both rifles and belt guns, so that our ammunition would be uniform, and, too, each had a saddle and packhorse. Two of the boys had gone on ahead to map out the route, and we were to meet them at Nogales to cross the border. However, on reaching Nogales, they were informed that they would not be permitted to take in guns and ammunition, and recommended that we call it off, as none of us wanted to be in Yaqui country unarmed.

While we were being lazy, reading, and enjoying the

sun, we noticed that our neighbor across the creek, who was the owner of one of the largest saloons in town, had returned from the outside, accompanied by a beautiful young woman. In those days, with everything running wide open, no one knew, or cared, whether she was his daughter, wife, or someone he had brought in to live with, so they were usually given the benefit of the doubt. One day, as she was sitting outside reading, the boys dared me to go over and get acquainted. I wanted to meet her anyway, so I crossed the footbridge and introduced myself. At close range she was even more beautiful than she appeared from a distance, deep auburn hair, hazel eyes, and the beautiful peaches-and-cream complexion that usually accompanied that combination. I found that she was his daughter, had traveled extensively with her mother before she had died, and had been educated in the best schools in the East.

We seemed to enjoy each other's company from the start, and, as we both had a number of good books, we exchanged them and spent considerable time together. This had been going on for some time when her father came home for lunch one day while I was there. He seemed a little upset when he called her into the house, but I thought nothing of it and returned to my cabin. After lunch he came over to where I was sitting, and I knew he had something on his mind, the way he kept standing first on one foot and then on the other. Finally, he said, "Say, young fellow, I don't have a thing in the world against you, but I would rather you stayed on your side of the creek." In those days a man's home was really his castle, and I thought I had better comply, but a few days later, while I was washing some shirts, she called across saying, "I have an iron if you want to iron those shirts." With this invitation our little visits started again, although we had to be discreet, so I just went over evenings when we could shine a large reflector-type lamp down the narrow trail on which her father came home, and, of course, I had always gone when he arrived.

At some time during our visits, she had told me that she was to receive an inheritance, left by her mother, when she came of age. I never knew how much, or the reason for her father's behavior toward her, until years later when a mutual friend told me the story. It seems her father had gone to the Klondike in the early days and made a huge strike, half of which he gave to his wife. His was promptly spent, but she used hers to better advantage, and, when she died, left it to her daughter. That was why her father did not want her to become interested in anyone as he wanted to keep her single until she became of age, thinking he could get control of her inheritance which was, according to rumor, around $200,000. I hoped the rumor was true, as she was a wonderful girl and I could never wish her anything but the very best.

Time went by all too fast, and it was the Fourth of July when I got two surprises, one of which deflated my ego considerably. When time came for the hundred-yard dash I was to run against Joe O'Brien, a mining engineer, and, I understand, a pretty fast man. All we could find to run in was tennis shoes. We were each watching the other, when out walked a huge man wearing running spikes. He was not fat, just big, and he had been drinking, so we both tossed him off as nothing to worry about. Well, he beat us both going away, and he was a picture to behold in action.

The other surprise was when my partner, as the winner of the high jump was about to be announced, casually trotted in from the side and cleared the bar. He had never told me that he was an athlete, but he could surely high jump, and was, in fact, a very good all-around man. Midnight did relieve my embarrassment somewhat by winning everything in which he was entered and made my day more bearable for me.

After the Fourth, we decided to go to the Matterhorn, where our claim was located, and do our assessment work, which was required every year for five years, after which the mine became patentable. While there, we decided to prospect some more and did find enough showings to justify holding onto the claim for future possibilities. This done, we returned to town, and decided to go outside for the winter. After making everything shipshape, and hanging bedding and food from the ridgepole with baling wire to protect it from the rats, we locked up and started out by way of Three Creek, Rogerson, Oakley, and Burley, from where we followed the Snake River to Idaho Falls and the sugar factory. Here we knew quite a few people, and usually enjoyed ourselves at the dances and shows. Toward the end of the run at the factory, my partner received a letter asking him to come home, as it was necessary because of illness in the family, and, as we had another contract with the Pavlak Mill to furnish cordwood, he asked my old friend Dewey, who was with us at the factory, to fill in for him. This made Dewey very happy, as he wanted to go anyway, so, as soon as the run was over, we saddled our horses and started back to camp.

By this time there were wagon roads to Jarbidge from both the Idaho and Nevada sides, and a considerable number of freighters were hauling supplies. One of them had a jerk-line team of twenty-two mules, and I liked to watch them negotiate a hairpin turn on the ridge above Pavlak, as they made the turn, one mule on each swing team would jump over the chain, so that he could keep pulling until the turn was made, and, at that time, he would jump back to his own side of the chain. These mules were trained to perfection, and I never saw them make a mistake.

This particular hill has other memories for me. On one trip I was making with a load of mill timbers, the rough lock chain broke about halfway down the hill. I heard the driver of the wagon behind me yell, "Jump!" but I could not bring myself to do so, as the four horses I was driving belonged to an elderly man in town, and they were all he had. At a time like this, it seems, a thousand thoughts rush through one's mind in a split second. I could not use my brakes, as they would go out long before I reached the bottom of the grade, which was the reason for the rough lock in the first place. If I abandoned the team, they would be out of control and most certainly would be pulled off the grade by the heavy load, in which case they would all be killed. The grade ahead was clear, my load was on a low-wheeled wagon with wide bunks extending out beyond the wheels, and was well bound with strong chains. The mountain was so steep that, in order to get a wagon road of sufficient width, a vertical cut of twelve feet was required. This proved to be a blessing in disguise because, by holding my team tight against this face, and gouging into it, I was able to check my speed somewhat. The leaders had to be kept running on a tight line so that the wheelers would not step over their single-trees, and pile up. At the bottom of the grade was a fairly sharp turn into a flat meadow, and I felt sure that we would roll over at this point, unless my speed could be further checked, but, at any rate, the teams would be on flat ground and have a chance. When we reached the turn, I had, by consistent gouging of the bank, been able to hold the speed down so that we stayed right side up, and by now applying the brakes, brought them to a stop. The first man over to me when we came to a stop said, "My God, man, you are either a damn fool, or you have a lot of guts."

He was wrong both times, as I thought I had used my head, and I had not had time to think or find out whether or not I had guts, but now reaction had set in, and I was scared to death. Fortunately for me, he could not see my knees bumping together, so he did not realize how scared I really was.

In addition to the horse-drawn freighters and stage-coaches, a few automobiles were now coming through, but the only ones able to successfully negotiate the grade out of the canyon were the White and Stanley Steamers, and their whistles could be heard for miles as they came winding up through the mountain.

With the improvement in roads and transportation, heavier equipment was being brought in, and most of the promising strikes had been taken over by the big financial interests, resulting in the town settling down. Mike Pavlak had sold, and, as I remember, received about $65,000 in cash and a block of stock in the company. He married a widow to whom he was reported to have given half of the cash, and proceeded to put on a big wedding party with champagne served from an elaborately decorated galvanized washtub. Everyone seemed to think she married him for his money, but, if she did, she kept her end of the bargain, as the last I heard of them she had bought a ranch for her son to run, and was taking good care of old Mike.

As a result of the town settling down, a little social life was beginning to appear, and we all got together to build a community hall. The logs we brought in from the surrounding timber, and the maple floors, doors, windows, and such were shipped in from Twin Falls. We were nearing completion when we had a deadline to meet for a special dance which turned out to be somewhat hilarious. The dressing we had put on the floor had not completely dried, and, when we started to dance, it formed a gummy mess that made it almost impossible to move, let alone dance. We cleaned the mess up, applied a liberal dressing of cornmeal, which was danced on once, and then swept up. Then we applied the wax, and after being danced on the floor was so slippery that it was difficult to stand up. Dancing at the time was a very small man with a partner roughly twice his size who was having difficulty. She finally fell, and, when she did, landed right in the little fellow's breadbasket. Showing no concern whatever for her partner, and with her ample posterior almost hiding the poor guy, she nonchalantly inquired, "Can anyone tell me exactly how I did that?"

Now that the excitement of the wedding and the first dance was over, we had to think about getting up to our cabin in the timber, and the completion of our contract with the Pavlak Mill. Dewey was thoroughly enjoying himself in this new environment, and he was an inexhaustible worker, so we made good headway and had everything finished ahead of schedule.

As it was then about time to do our assessment work on the claim if we were to beat the snow deadline, we decided to go up and get it over with. It was here on this high mountain, at about 10,500 feet elevation, that I met my Waterloo. We were driving a tunnel to cut a surface vein we had discovered, and the ground was very unstable, giving us considerable trouble trying to get started under. An open cut twenty-two feet in length, with a twelve-foot face, had already been excavated, and, in trying to get under, I drove my pick into what appeared to be a small rock, but it must have been the key to the whole face because it suddenly caved in on me, and, being between the walls of the cut, I had no chance to escape.

Dewey, who was in the clear, started immediately to extricate me from my prison of rocks and debris, and managed to get a huge rock off my leg that two much larger men he had taken there a few days later were unable to move. This, I believe, tends to prove that in emergencies a man has great reserves of strength to call upon that he does not normally possess. At this point I was numb, and possibly somewhat in shock, since I felt no pain, but, with the removal of the rocks, I began to experience extreme pain and realized I was seriously hurt, how serious I was to find out later.

How Dewey ever managed to get me on my horse, and how I endured that eight miles to town I will never know, but do it we did, and, upon arrival, found that the only doctor in town was a retired army surgeon, which was possibly fortunate for me as he certainly had experience. After an examination, he informed me that I had a broken collarbone, two fractured ribs, and a leg broken in two places. There was no anesthetics to be had, so he gave me a stick to bite on, and two others to grip with my hands, while he fixed me up. It was about ninety days before I was able to get along by myself, and I realized it would be a long time before I would be able to roam these hills again with any degree of pleasure. This being the case, I decided to go out to Twin Falls.

CHAPTER X

Twin Falls

MY DECISION TO GO TO TWIN FALLS WAS DETERMINED
not only by my injuries but by the fact that my engineering course had been completed for some time and
I was looking for practical experience. All my studies
had, of course, been theoretical. Upon my arrival in
Twin Falls I was very fortunate in securing employment with a St. Louis firm buying and shipping alfalfa.
My work required some training, and traveling between
Twin Falls, Buhl, Filer, Kimberly, Hansen, and Burley,
and, as I had my own horse, Midnight, I was allowed
five dollars per diem for him, in addition to a good salary for myself. This was very good for me as it required very little physical effort, and allowed me to get
back in shape. We shipped about three hundred cars
that year, and filled every available warehouse in the
towns mentioned. In my spare time I looked for the
best machine shop in town, and found an excuse to get
acquainted with the owner, as I wanted to get some experience on the various machines. Having completed
the course in mechanical engineering, I was a fairly
good draftsman and felt that I could be of service
around a machine shop, which turned out to be true,

and the owner was as glad to have me around as I was to be there.

During 1912 I married, and, when I informed my mother, who had come to the United States during 1907, of the coming event, she replied in her inimitable manner, "Well, I hope you don't pick a crab apple after running all over the orchard."

My work with the St. Louis company, buying and shipping alfalfa, continued until the spring of 1916 when ground was broken for a new sugar factory, and, thinking I might get a job where my services could be used, I went out to see the superintendent and gave him my qualifications. He did not promise anything but took my name, address, and telephone number. As I was leaving he was having considerable trouble with his car, and had a mechanic working on it with no apparent results. It was an E.M.F. commonly known as the "every morning fixit car," built by Everett, Maxwell, and Flanders. Fortunately I was somewhat familiar with this make of car, and knew its peculiarities, so I was able to get it going, and, while the superintendent was noncommittal, I felt that I would be hearing from him.

I had lived in Twin Falls for the past three years, and my first child, a daughter, was born there on September 9, 1915. She came into the world under rather unusual circumstances, as we had lost our home and everything in it by fire, so she spent her first four or five months at the Rogerson Hotel, through the courtesy of Bob Rogerson, a good friend.

I remember that whenever he was introduced to anyone he always said, "Aye, Bob Rogerson, from Glaskee Mon."

My second child, a son, arrived thirteen months later on October 11, 1916. I enjoyed them as they were growing up, and we were quite close. They were both good students, and possessed the usual amount of humor inherent in most children. I remember when we decided to get rid of my son's Dutch cut hair and make him look more like a boy, which exposed the back of his head, my daughter came to me, and inquired, "Daddy, what are those bumps on the back of George's head?" To keep her quiet, I replied, "Oh, those are just signs of intelligence," only to be informed, "Well, Daddy, you don't have any," which stopped me cold.

They were also well disciplined, and seldom questioned an order, which was fortunate and paid off several years later in Montana. The rivers were at flood stage and all the wild life that normally lived along the banks were heading for high ground. We were watching the rising water, and my son, who was then nearing eight, was standing about five feet distant when I happened to glance back and saw a huge rattlesnake approaching from the rear. It was so close that there was only one thing to do and that was to remain absolutely motionless. I dared not raise my voice, so I said in a normal manner, "George, listen to me carefully. It's important. Don't ask me why, but don't move." Without question, he became rigid as a statue, and the rattler passed within six inches of his feet. Had he moved the snake would undoubtedly have struck. As soon as there was enough distance between them, I disposed of the menace.

There were quite a few Scottish sheepmen in the area of Twin Falls, two of whom were bachelors. The story goes that, on one of their trips East, one of the brothers brought a very attractive young woman back with him, and established her in an apartment as his niece. Some time later, during a poker game, the subject of the young lady came up, and the brother said, "Well, she may be a niece of Bob's, but she's no relative of mine!"

A few days after my interview with the superintendent at the sugar factory, I received a call to report for work. The company had contracted to build the factory and operate it the first year, as was the custom since there was a lot of machinery and steam fitting involved.

The first unit to be built was the machine shop, which consisted of lathes, shapers, milling machines, grinders, and pipe machines. As soon as the building was complete, and the machines arrived, I was ordered to set them up, and get them in operation, which required a considerable amount of millwright work. Upon completion, we started cutting various lengths and sizes of pipe nipples, as they would be required in great numbers once construction of the factory got under way. Everything up to ten inches in diameter was cut on the pipe machines, and, over that, on a huge lathe which was still there when I visited about five years ago. By this time the heavy equipment was arriving, huge Corliss steam engines, high-speed dynamo engines, and dynamos to generate power to run the plant, evaporators, crystalizers, brown and white pans, centrifugals, and all the necessary equipment for making sugar.

My past experience in the sugar factory was very valuable to me now, as I knew the workings of a mill from the time the beets came from the sheds, until the sugar was sacked, weighed, and stored. The lowly beet traveled a very circuitous route before it became refined for table use, and required a tremendous amount of machinery, piping, and pumps to carry the juice to the various stations.

All my hours of study through the long evenings were now paying off, as I had no difficulty reading and interpreting the various blueprints involved, or, if necessary, making a drawing of any special part required.

The factory was completed on schedule, and we started our first-year shakedown run to correct any bugs that might develop. As expected, the usual difficulties were encountered, and we lost two men before the run was over, one of them because of the carelessness of the victim himself.

Being the first year, the run was shorter than usual, as the farmers had not planted to their full capacity. My boss, Mr. McLaughlin, was the typical hard-boiled superintendent, but we had always worked well together, so he asked me to stay with him on future jobs. However, I had dreams of greater things, as I had made good use of the machine shop, and had developed several ideas I had been thinking of for some time.

I loved experimental work, and had already built several devices which had proven to be successful, but that I had never bothered to patent. Among them were an automatic device to fire a cartridge at predetermined intervals which we used on the lambing ground to keep the coyotes away; an electric gaff for loading cattle, as I thought the old way was much too cruel; and an electric mouse or rat trap. I believe the device for loading cattle is still being used at all the major loading corrals or yards. Now, however, my doctor, who was quite well to do, was financing two of my ideas, one of them being a multi-gap spark plug, as we had so much trouble with fouled spark plugs in those days. My patent on this was basic and covered an interrupted spark gap, which, I was told later, was holding up some of the things being done in relation to X ray. My doctor, being interested in those things, talked me into waiving my patent rights, "for the good of humanity," which I did, and we thereby both lost a fortune. We did go ahead with the manufacture of the spark plugs, getting our porcelains from France, and were fortunate in developing a good market.

My other idea, a working model of which had already been built, was a rotary induction and exhaust system for internal-combustion engines. It was accepted by most everyone as being feasible, and a corporation was formed for the purpose of raising the necessary financing to build a full-size cylinder head, which was to be made in Detroit under my supervision. This being the case, certain arrangements had to be made, as I did not know how long it would be before I came back to Twin Falls.

My horse, Midnight, was a problem. He was now seventeen years of age, and we had been over a lot of country together. In trying to find a solution, I remembered that George Stanger, his original owner, had once said, "If you ever want to sell him, let me know." I sent him a letter explaining that I did not want to sell, but that I did want to know that he had a good home, which I was sure he would have at his ranch. After several letters, we arranged to meet at American Falls, where he was to take possession, so I shipped my things to Detroit, and rode horseback to American Falls. When we met, Mr. Stanger was accompanied by his friend, Valdez, and they both seemed very glad to see me again. We spent an enjoyable evening together at which he assured me that both horse and equipment would be well taken care of, and available if I needed them. As time grew near for my departure, we went to the depot, all feeling a little sad, I thought, at our parting. They thought I should not be giving up the things I loved so much, and I tried to convince them that I also loved to be creative, and had spent many weary hours preparing myself for this. I was, of course, unhappy about leaving Midnight, even though I knew he would have the best of care, as I was saying good-bye to an old and trusted friend. Through the years that followed I wrote to my friend, Stanger, and he informed me that for about a year Midnight was not interested in anyone or anything, but he finally warmed up to him a little, and they became good friends. He said that he rode him quite often, and, when he did, always used my black outfit as they seemed to belong together. He also said that both he and Valdez, when their families got together, and the youngsters asked about the black saddle, which was quite unusual, enjoyed telling them about the English boy who came riding in on a big black horse that they had given up as an incorrigible outlaw.

CHAPTER XI

Detroit and Automotive Engineering

UPON MY ARRIVAL IN DETROIT, I FOUND A ROOM WITH a very nice family, which was much better than a hotel as I wanted a quiet place to work. My family had remained in Twin Falls, as my time in Detroit was somewhat uncertain, and I did not want to be moving them from place to place. Now, being settled, I purchased a drawing board and started to prepare my working drawings, and to look for a contract machine shop where the work could be done. Here, too, I was fortunate in finding a shop owned by two brothers. They were Swiss and excellent machinists, as most Swiss are, so I really had no problem, and, in about sixty days, we had an experimental cylinder head completed and working. It was at this point that my troubles began. My company in Twin Falls suggested that since it appeared I might be located in Detroit for some time, I have my family move there, which I did. About thirty days later the blow came, and I realized that I had been very naïve in my business transactions. I had not only given them control of the company, but had also signed an agreement, which I should have read more carefully, giving them control of anything I might in the future invent. This was naturally a shock to me, as I was already working on several other devices, and, under the circumstances, I certainly did not want them to benefit. They had already stopped my salary, as a preliminary to their freezeout, and several weeks later, I received an offer of $2,500 for my block of stock. This, I thought, was too little, but I would have given it to them to get back the contract I had signed, and, with this in mind, I advised them that I would accept their offer, provided it was accompanied by my cancelled contracts, to which they agreed. Obviously, in their anxiety to obtain complete ownership, they, too, had overlooked the well-hidden clause.

While this was going on, the war in Europe was coming ever closer to the United States. I had not been too interested, as I had been in the States for several years, and, of course, felt that this was now my country. I had followed the news closely, and quite naturally was sympathetic toward my old home, England. Being a student of aeronautics, I had followed closely the activities of the various aces, and the planes of Britain, France, and Germany. I had read of the performances of the Bristol Bullet, Sopwith Pup, and the FE2D of Britain, the Nieuport and Spad of France, and the Albatross and Halberstadt of Germany. I had also read and studied everything I could find relative to the aircraft engine, and was quite familiar with the Gnome, Le Rhone, Hispano-Suiza, Fiat, Sunbeam, Rolls-Royce, and Mercedes engines. With this background, it is not surprising that I leaned toward the aircraft section of the armed forces, so when the Germans bombed a town in England I knew quite well, a town that had no military value whatever,

I was furious, and promptly left for Windsor, Canada, to try and enlist in the Royal Canadian Air Force. I passed the preliminary examination and about one week later received a telegram ordering me to report at Toronto. Here, because of my accident several years before, I was unable to pass the physical test, and returned home to Detroit. Soon after this, in April, 1917, the United States entered the war and the draft was instituted. As I was married and had two children, my classification was 4B, but a short time later, when men were called for, I reported to the Grace Hospital for examination, only to fail again because of my prior injuries. At that time the commanding officer informed me that I would probably hear from them, as men with my qualifications were badly needed. Things were now beginning to hum and two automotive engineers were assigned to design a suitable aircraft engine. These men, Jessie G. Vincent and J. G. Hall, went to work and about ninety days after they started, the first engines were ready for testing. The first were eight cylinder, designed to deliver up to three hundred horsepower, but things were moving so fast, that, before they got into production, a more powerful engine was needed, and the engineers started the design of a twelve-cylinder engine of four hundred horsepower. I was happy to be assigned to this project and enjoyed my work very much.

The first cylinders were hogged out of solid steel billets. Then Harold Wills, who later built the Wills St. Claire automobile, but was with Ford at the time, developed a method of making cylinders out of steel tubing. This method was so successful that the Ford Motor Company made all the cylinders used in the Liberty engines, and was, as a result, the last to go into production on the engine itself. Others manufacturing the Liberty were Buick, Cadillac, Marmon, and the Lincoln Company, which was formed for the express purpose of building Liberty engines. Like all other engines, the Liberty had its peculiarities. It had a bad habit of back-firing through the carburetor, creating a fire hazard. It also had a peculiar roll at low throttle, because of the fact that, instead of the cylinders being set at sixty degrees, which would be normal for a twelve, they were set at forty-five degrees to cut down frontal area, consequently, the firing order, instead of being at sixty-degree intervals, was forty-five degrees and seventy-five degrees, thereby creating the roll.

When we entered the war we had no planes of our own, at least not war planes, so it was decided to use the British Bristol and De Haviland, and the Italian Caproni. I understand a few Bristols were built, although I never saw any. We did build quite a lot of De Havilands and some Capronis. In the Detroit area these were built by the Fisher Body Company, and, when they were ready for the engines, I transferred and worked on the final test of the De Havilands. There were five of us assigned to this project, all of us physical rejects from what was then the "Aviation Section of the Signal Corps." However, we all had backgrounds of automotive engineering, which was so necessary at that time. Our responsibility was the final checking and adjustment of all equipment, including guns, before flight tests. At that time qualified personnel was in such short supply that only one out of four planes was flight-tested before being shipped overseas, and the engineer certifying the plane for flight accompanied the pilot and rode in the observer's seat. Consequently, even though we were not pilots, we did quite a lot of flying, as the overworked pilots were only too glad to get a little rest while in the air. It has been my observation that we, as Americans, have the fault of not being able to accept the fact that some other country may be able to do a job as well, and sometimes better, than we do. This was particularly true of the De Haviland. The British De Haviland had the engine up front, then the gas tank, and behind this the pilot and observer, so that, in case of a crash, there was no weighty material back of the men. Then, too, it was easy for the men to converse as they were next to each other. The American De Haviland had a Liberty engine in front, then the pilot, followed by a ninety-eight-gallon fuel tank, and behind this the observer. Obviously, conversation between pilot and observer was not as good with a tank of this size between them, and, in the event of a crash, the pilot simply became the filling of a sandwich between about one thousand pounds of engine and ninety-eight gallons of gas. In spite of all this, the five of us checked out over two thousand of the planes without a serious injury.

We did, however, have some very close calls. I remember one day a cover of an electric junction box came loose, and the propeller picked it up and drove it through a protecting steel screen, where it cut a swath across the stomach of one of my assistants. It took a slice out of his overalls, trousers, and underwear, scratching his stomach quite deeply. An inch closer, and he would have been disemboweled. On another occasion, I was checking the guns for synchronization with the propeller, which was the last test we usually made. Our specifications called for a burst at three hundred revolutions per minute, another at one thousand revolutions per minute, and the last burst at seventeen hundred revolutions per minute, or full throttle. I had fired bursts at the lower speeds, but when I squeezed the gun control at full throttle all hell broke loose. I could hear something screaming over my head but did not realize that it was the bullets from my own guns. What we had all overlooked, was the fact that we had been shooting about sixty-four hundred rounds of ammunition into the sandbank each day for the past two weeks when this happened, and the lead had all accumulated into large masses. My last burst had ricocheted off one of these, some of the bullets going over the bunker and across a large assembly room, others making a 180-degree turn right back over my head. My guardian angel must have been watching over me that day, because, had I been sitting up in a normal position, the bullets would have taken my head off right above the eyes. As it was I was leaning forward, checking some of the instruments, so the bullets just missed me. After this accident, the sand was screened every weekend.

The rest of the boys had their close calls, too, but we all came through in one piece. There were, of course, our lighter moments, and, in retrospect, I suppose we were, at times, bad boys. Certainly we had little respect for anyone, and one of our favorite stunts, when we had celebrities visiting, was to get the ladies in the proper position, with the engine idling, and, at a given signal the throttle would be shoved forward giving them the benefit of the propeller wash, and, of course, making it quite difficult for them to keep their dresses in the proper place. All of us apologized very profusely, with tongue in cheek, so we seldom received a reprimand.

One other incident that was amusing to me was on Armistice Day, when most of the girls, before they left, wanted to know how it felt to sit in the cockpit of an airplane. These girls were from all walks of life, and the questions they asked relative to planes and flying were most amusing.

Later on Armistice Day, we had a tragedy when Lieutenant Morrow, flying too low, brushed a flagpole, damaging one of his ailerons and throwing him out of control. He never had a chance, and when he crashed, with only the four one-and-a-half-inch spruce longerons holding the engine and fuel tank apart, he was crushed flat. The observer suffered a broken leg and minor injuries, as there was no weight behind him.

After the Armistice, when most everyone was looking for any available job there might be, Charles Bradford and I, who were the first on the job, were asked to stay until everything was cleaned up. We were assured that if we did we would be taken care of when the time came. I was very glad that I decided to stay, as it involved some new and very interesting experiences, among them, working with a group of Italian engineers who had come to the United States to build Caproni planes. This had been a hush-hush project and very few people knew anything about it. However, they had almost completed three huge triplanes of one hundred forty feet span, the top wing being thirty feet off the ground, and were to have been powered with Fiat engines. As the Liberty engines were now available, it was decided to use those, and we were assigned to help them with the conversion. In spite of the language barrier, we enjoyed working with them as they were all master craftsmen. Their commanding officer was a very pleasant young man who used perfect English which was understandable, as he was, I believe, related to Gabriele d'Annunzio, one of the great novelists, dramatists, and poets of Italy. The Caproni was a huge plane, using three Liberty engines for power, two tractors, and a pusher. The two tractor engines were mounted in outboard nacelles, with the propellers in direct line with the pilot and copilot, and, as it was an open cockpit, as they all were in those days, I have to admit to a little nervousness, sitting there so close to those propeller tips. Whether or not those Caproni's ever saw service I do not know, but we did complete three of them, and Bradford and I attended a farewell party for the Italians when they returned to their homes.

It was almost a year after the Armistice before we were finished, and, true to their word, both Bradford and I received a call with an offer of a good position. He went with the Franklin Automobile Sales and Service, and I received a call from the president of the Detroit Creamery. They operated a fleet of about four hundred trucks of all sizes from the ten-ton Macks to Fork pick-ups, and about half of them were inoperative, as they had lost most of their good mechanics in the war. He asked me to take over and straighten things out, and to name my own salary. This was, in a way, a challenge, so I accepted and went to work. It required about eight months to get things running smoothly, and to train mechanics to replace those who were lost.

This accomplished, I grew restless as there were things I wanted to do, and the only way it could be done was to be in business for myself. The war's end had seen great advances made in aircraft design, due to the competitive urge of hostilities, but this era was closed by the cancellation of orders for military aircraft before a civil aviation had been established. Companies who had met the ever-pressing demand for aircraft during the years of conflict were hard hit, and many closed down, not to reopen for many years, if at all. So the aeronautical field, where I wanted to work, did not look too encouraging. At this time one of my coworkers on the De Haviland project, Ed Weeks, who had been a flyer before the war, wrote me relative to starting a flying school. He had learned that one hundred Jennies in their original crates were available, and, as these planes had been the standard United States training plane, he thought they would be ideal for school purposes. We investigated and found that the price was eight thousand dollars each. Later we learned that they all had been bought for much less than the quoted price. Obviously, the purchaser knew someone in high places. This, of course, ended our dreams of a flying school, and our financial backer was just as disappointed as we were. In the postwar years, aeronautics was not the only industry in the doldrums. Business generally was in a slump, so I decided the only thing to do was to cater to those with the ability to pay for the services they required, and, with this in mind, I rented a building just completed and opened a service for Cadillac and Franklin cars. This venture was not a huge success but it did bring in sufficient income to enable me to continue with my experimental work.

During the war I had become quite interested in the problem of power loss at the higher altitudes, and, since the power output of an internal-combustion engine depends on its breathing capacity, my research was directed toward the problem of compensating for the loss of pressure in the higher altitudes. Very little, if anything, had been done in this field, so I decided to explore the possibilities of superchargers as a means of forcing more air into the cylinders. This, together with the redesigning of the induction and exhaust systems, showed a marked increase in power output, so we were quite satisfied with our progress. However, conditions were slowly going from bad to worse, and many of the automobile companies were going out of business. A few were hanging on, and, in 1921, one new company was formed using the famous name of Captain Eddie Rickenbacker and his hat in the ring insignia around which to build a car. In 1922, the Lincoln Motor Company, built by Henry M. Leland to manufacture Liberty aircraft engines, was bankrupt, and was bought by Henry Ford. Walter Chrysler took over the Maxwell Chalmers Company during 1923 and started the Chrysler line.

During this same period, aviation was kept alive by a few who had faith, and, I believe, one of the contributing factors was the Schneider Trophy Races. They did as much, or more, than anything else at that time to further aerodynamics, improve aircraft structures, and to advance the speed of flight.

The trophy was won in 1922 by Captain Henry C. Baird flying a supermarine powered by a 450 h.p. Napier Lion engine, for Britain at 145.7 mph. Lieutenant David Rittenhouse, flying a Curtiss Navy Racer, powered by a 465 h.p. Curtis D12 engine, won at 177.38 mph for the United States in 1923. One of the observers at this race was Dick Fairy, of England. He was deeply impressed, both by the D12 engine and the aluminum alloy propeller that was then coming into use in the United States and obtained manufacturing rights for both.

About this time, my neighbor, who occupied the adjoining apartment in a four-family flat, had a visitor from Montana. He turned out to be the Ford dealer for Miles City, and, after some discussion as to the economic conditions in various parts of the country, he informed me that things were much better in the West, and that with my background I could no doubt do very well there, as they were sending all their electrical work, generators, starters, and magnetos to Minneapolis for repairs. He wanted to see my shop, which was equipped to do the work they were sending out, and the next day, after looking it over, was sufficiently impressed to urge me to go to Montana. This was quite a move for me so I told him I would have to be assured of support before considering it, so, upon his return to Montana, he talked to some of the other men and I received a letter indicating their support, so decided to take the gamble, and, buying a truck in which to haul my equipment, started preparations for the trip. One of the men, a Pennsylvania Dutch boy, had saved a few hundred dollars and wanted to go West, so, as he was a fine mechanic and auto-electrician, I was very glad to have him, knowing that such men would be hard to find in Montana.

CHAPTER XII

Miles City, Montana

As we had worked through the war years without time off, it was decided to make a vacation out of
this trip and take it in easy stages, camping along the
way. Leaving Detroit, we crossed southern Michigan
to Chicago, and trailed along the west shore of Lake
Michigan to Milwaukee. From there we traveled northwest, across Wisconsin to Eau Claire and St. Paul, and
then, still northwest, across Minnesota to Fargo, North
Dakota. At that time of the year, Wisconsin and Minnesota were beautiful and the fishing was excellent, so we
were certainly having an enjoyable trip. From Fargo
west across North Dakota to Jamestown, Bismarck, and
Miles City, Montana, the country was more open range,
but still enjoyable, although we missed the wooded country through which we had come. Arriving at Miles City
(the home town of Fort Keogh, from which, according
to the old-timers, General George A. Custer started up
the Yellowstone and Rosebud to keep his appointment
with destiny at the "Battle of the Little Big Horn River"), we found a building, moved in, and started to set
up our equipment. We had excellent support and were
soon servicing practically all the automotive electrical
equipment within a radius of seventy-five miles. How-

ever, the fact that I had left Detroit did not mean that I had severed connections, or that any of my ideas had been abandoned. I still had a machine shop, even though small, with which to work, and had maintained all my technical contacts and papers, including those from the National Advisory Committee for Aeronautics. My dream of building the perfect aircraft engine was still with me, and I was working on my drawing board to that end, as well as keeping in close touch with developments, particularly the Schneider Trophy Races. So I was deeply interested when Lieutenant James Doolittle, flying a Curtiss Army Racer, powered by a Curtiss Y-1400 of 600 h.p. won for the United States at 232.57 mph in 1925, and when, in November of 1926, Major Mario de Bernardo, flying a Macchi Moro Seaplane powered by a Fiat engine of 800 h.p. took the trophy back to Italy with a record of 246.442 mph. Now, as a result of the Schneider Trophy Races, aeronautical engineers in all major countries were becoming aware of their technical importance. This was particularly true of Great Britain, and the Air Ministry established an experimental High Speed Flight Station. Candidates for high-speed flying were drawn from the elite of the service test pilots, who would eventually challenge all comers for the Trophy. Consequently, during 1927, at Lido, Italy, Flight Lieutenant S. N. Webster piloted an S5 Low Wing Monoplane powered by a Napier Lion engine of 875 h.p. to win at 281.65 mph. By this time a set of working drawings were almost completed incorporating some of the ideas I had in mind, including the maintenance of power at altitude.

That same year, Lieutenant James H. Doolittle and a group of flyers set down on the flat above Miles City. I met him at a dinner the town gave for them, and, while I never became too well acquainted with him, through the years I always watched his activities from afar. He is, I believe, together with Ernst Udet and R. L. Atcherly, one of the most natural flyers of all time. He seemed to do the right thing at the right time always. It was these men, and others of their type, as well as the engineers who designed their planes, that kept aviation alive throughout the lean years. It was also during 1927 that Charles Lindbergh gave aviation a much-needed boost by taking off in his *Spirit of St. Louis* from Roosevelt Field to cross the Atlantic. A lot of prayers were riding with him, but very few thought he would make it, and, when he landed at Le Bourget, in Paris, a sigh of relief was heard around the world.

After Lindbergh's flight most of the businessmen of Miles City were pressing me to get started on my engine. There had been several pilots and engineers stop by to see the model I had built, and they were all enthused about its possibilities. Actually I was not quite ready, however, though this was very unusual, I was constantly under pressure to begin. Finally I told them I would have to have sufficient money in the bank to complete the project before starting, and, not long after this, I was informed that the money was in the bank and that they would like to incorporate and get started. While the attorney was getting the corporation set up, I received a letter from E. E. Porterfield, Jr., president of the American Eagle Aircraft Corporation, of Kansas

City, inviting me to come there and use his facilities. This letter was presented at our first Board of Directors meeting, and it was decided to accept his offer. Consequently, in June of 1928, I packed my car and started East to build my engine.

CHAPTER XIII

The Franklin Aircraft Engine

UPON MY ARRIVAL AT THE AMERICAN EAGLE PLANT, I found that their ideas of the equipment required to build an aircraft engine were quite different from mine. They did have a well-equipped plant for the building of airplane frames, fuselage and wings, but nothing that could be used in the building of an engine. I should have realized this, as I knew their power plants were obtained from engine manufacturers. However, they were helpful in locating a contract shop with the equipment and ability to do the work, and arrangements were made for the machining of the various parts. It was also necessary to find a competent patternmaker and foundry, as these, of course, came before the machine work. My greatest problem was the machining of the cam to be used, as very few shops were set up for this kind of work. Finally I sent blueprints to the Rowbottom Machine Company at Waterbury, Connecticut, specialists in this kind of work, and they made them for me. With the preliminaries out of the way, I settled down to a general supervision of the job as it went through the various operations, and, in about ten months, had an engine ready for dynamometer tests. We were all pleased with the results, as were all the engineers, American and foreign, who witnessed its performance, but there was still a lot of work to do, checking and double-checking the various parts for wear, gas consumption, lubrication, and other things. While this was going on, I was trying to keep up with the developments going on around me, both in the automotive and aeronautical field in 1927. I was happy to see the low-wing monoplane emerging as the favorite speed plane, as I had done considerable experimental work with flying models several years before, only to be laughed out of my own shop by my contemporaries. This, then, was vindication.

During 1929, I was intensely interested when Italy challenged the British at Spithead, only to be beaten by Flying Officer H. R. Waghorn, flying an S6 powered by a nineteen hundred horsepower Rolls-Royce engine at 328.63 mph. A few days later, Squadron Leader A. H. Orlebar flew to a world record of 357.7 mph.

All through this period we had been testing, disassembling, checking, and reassembling to run again, until we had accumulated about two hundred hours of running time, and were ready to proceed when the bottom dropped out of everything in the big financial crash. After this it was impossible to finance anything as there simply was no money available. This was a bitter disappointment to me as we had completed all our tests and were ready to go. However, there was nothing to do but keep trying. We rented an office in the Fairfax Airport Administration Building to be nearer what little activities there were, and I did get a Dr. Cross interested. Dr. Cross was a scientist, and reported to have an enormous income from processes he had developed. He had his own plane and pilot, as well as several engineers in his employ, and all were enthused with the possibilities to the point where he was anxious to go ahead. Lady Luck, however, was not with us. Dr. Cross loaned his plane and pilot to four of his friends for a fishing trip to Corpus Christi, Texas, and, on the way home they crashed, killing them all. Upon receiving the news, Dr. Cross was put to bed by his doctor and within the week died, taking my dreams with him. This accident left us all very depressed, but we had to carry on, and, even though conditions were very bad, a few others, too, were trying to pick up the pieces and go ahead.

One of these was the Travelair Company that came out with their mystery plane about this time. These planes had excellent performance characteristics, and were flown by Jimmy Doolittle, who was aeronautical advisor to the Shell Oil Company at that time, and Frank Hawks, who had a similar position with Texaco. They both did a beautiful job putting these planes through their paces, and, as I remember, they outperformed all the service planes.

In 1931, Britain was also having problems, and, although another win meant permanent possession of the Schneider Trophy, it was decided not to underwrite the race. However, just when things looked darkest, Lady Houston agreed to underwrite the race enabling them to build two S6B planes powered with Rolls-Royce engines of 2,300 horsepower.

There were no challengers, but Flight Lieutenant J. N. Boothman flew the course at 340.08 mph to a technical victory, and won by default, thereby insuring permanent possession of the trophy by Great Britain. The Sister S6B, with its engine boosted to 2,600 horsepower, later that year won the world speed record for Britain at 407.5 mph. I remember discussing this performance with a representative of the Fairy Company, who was in this country, as the technical data indicated a speed higher than the rpm of the engine and the pitch of the propeller warranted. We finally came to the conclusion that the metal propeller, which was geared to half engine speed, and designed for a heavy pitch, could have flattened slightly, thereby increasing the pitch and speed.

In late 1931 I received a letter from a Mr. Uppercu, who was the Cadillac and LaSalle dealer for the whole metropolitan area around New York. How he heard of me I do not know, but he wanted me to bring the engine to New York, and discuss it with his engineers at the Aeromarine Klem Corporation, Keyport, New Jersey, of which he was president.

We had lunch at the New York Athletic Club, and he told me that they had lost 60 percent of their membership because of the depression. After discussing the engine for several hours, his engineers agreed that it had many possibilities, and he wanted to call Captain Rickenbacker, who was then, I believe, president of Eastern Airlines. He was, of course, a very busy man, and did not want to take the time, but Mr. Uppercu finally talked him into coming for five minutes, if he would pick him up and take him back. I felt sure he expected to see a hodgepodge of metal, as most experimental jobs were, but when I took the cover off, he said, "Well, you must have spent a lot of money on this," and visited for an hour before leaving.

I also had the pleasure of meeting Alexander P. de Seversky, who came to the United States in 1918 after the Russian Revolution. He was, during the thirties, considered one of the most advanced aeronautical engineers, and was responsible for many of our fighter planes. Later he wrote *Victory Through Air Power*.

Another pleasant surprise, while in New York, was meeting my old friend, Captain Frank Hawks. I had supposed that, since he was with the Texas Company, his office would be in Texas, but he actually occupied a good portion of the Chrysler Building's thirty-sixth floor. It was ironical that, with all his experience as a flyer, he was killed a few years later in a plane that was supposed to be safe for a novice.

In spite of the fact that engineers, both American and foreign, endorsed the engine, finances were nonexistent, very few people had any money, and the ones who did were certainly hanging onto it. They were hurt, and afraid to venture again.

I did have several offers, but they all wanted to control the corporation, which I found was the usual thing when the big boys were involved. Then, too, I suppose they thought I would meet any terms to get started, but having had one bitter experience I was not inclined to try it a second time, and I decided to try it my way. So with things getting progressively worse in the eastern part of the country, and remembering that once before

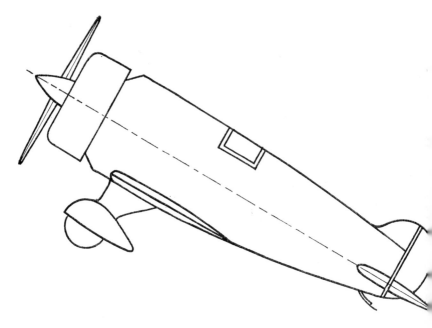

SKETCH OF AIRPLANE DESIGNED AROUND THE
CONVENTIONAL ENGINE

Note the large frontal areas and the amount of the propeller that is blanketed by the engine. This is, naturally, not conducive to high-speed flying where impact pressures are a problem for any designer of high-speed craft, and yet it is the best we have today and is the type of airplane flown by Frank Hawks, Jimmy Doolittle, Lowell Bayles, Weddel, and others. It is not hard to realize that a plane designed around the Franklin engine would have better performance in every respect. Drawings by G. E. Franklin.

PLATE XVI

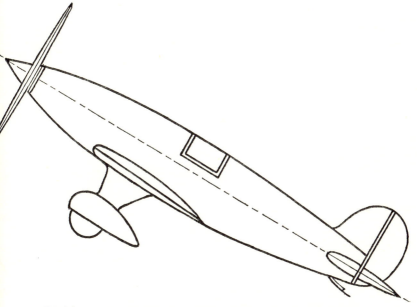

SKETCH OF AIRPLANE DESIGNED AROUND THE FRANKLIN CAM ENGINE

Attention is called to the exceptionally small frontal area and to the clean design which is impossible of attainment with the conventional engine. For military use this engine is unexcelled as the gun may be fired through the shaft, thus eliminating the possibility of damaging the propeller blades, as sometimes happens with synchronized guns.

PLATE XVII

I had found conditions better in the West, I turned west again, this time to Reno, Nevada.

For the benefit of the technically-minded reader, I am incorporating a brief description of the engine, together with photographs and reproductions of some of the working drawings, as well as photographs of the engine exhibited as part of the Aviation History Project, Montana Historical Society, Helena, Montana.

The Franklin engine is not just another engine haphazardly thrown together; it is the result of many years of conscientious effort and experience. Fortunately, it was my privilege to make an exhaustive study of the Almen engine built by the Experimental Station, Wright Field, Dayton, Ohio. This engine had many desirable features; unfortunately, there were difficulties involved that could not be overcome.

Knowing the inherent weaknesses of this particular engine, I set to work to overcome them and still retain those features that were sound. Both the Almen and the Mitchell operate on an inclined disk, or wobble plate, with a frictional element operating against it. In the Almen, this gave considerable trouble at high bearing pressures from lack of lubrication and while the Mitchell was improved in this respect, I felt that it would also give trouble at the higher bearing pressures. Consequently I used a cylindrical cam, hardened and ground, with a cam follower consisting of a roller mounted on ball bearings. This entirely eliminates the lubrication problem and insures a unit that will stand hard and continuous service.

The engine is of the barrel type with cylinders arranged in opposed pairs distributed cylindrically about the drive shaft, with their axes parallel to that of the drive shaft. The action of the engine is generally similar to that of other barrel types, except, that the drive is through a cylindrical cam instead of through the more common "wobble plate." Attached to each pair of pistons is a cam follower consisting of a roller mounted on ball bearings, a special bearing is required as the ball retainer is subject to reciprocating forces that do not occur in the usual bearing application. It is, therefore, very important to have a sturdy lightweight retainer supported on the outside diameter of the inner race. Bearings used are a special precision with a solid machined retainer of a special grade of formica, developed in cooperation with the Marlin-Rockwell Corporation.

All of the followers lie in the groove of the cylindrical cam mentioned, or on each side of the cam if a male cam is used as in a later engine. This cam is in turn connected to the main drive shaft of the engine.

The engine has sixteen cylinders mounted in opposed pairs as described; operates on the two-cycle principle; and is air-cooled, having longitudinal fins cast on the cylinder walls and heads. Since it is recognized that both air- and water-cooled engines have their inherent weaknesses, the water cooled having plumbing problems, leaks, and hose connections, and the air, control of cooling areas, it was decided to use, in later models, a combination of air and liquid "ethylene glycol" which has a boiling point of approximately 360°, for cooling, the liquid being sealed in and using a small expansion tank with no radiator, the cooling fins being the primary source of cooling.

Fuel is supplied by a supercharger which is so designed as to give sea-level conditions at fifteen thousand feet altitude, and is of the low speed positive type. With this system, crank case compression and the mixture of the oil with the fuel is eliminated as the fuel is forced directly into the cylinder.

Cylinder dimensions of the present engine are $3\frac{1}{4}$x6", giving a total displacement for the sixteen cylinders of 796 cubic inches; the estimated horsepower of the present engine is 400 h.p. at 2,000 rpm. Weight, without starter and generator is 359 lbs. It is expected that in a more finished design, total weight, including starter and generator, will not exceed 400 lbs., i.e., 1 lb. per horsepower.

Cylinders are constructed of aluminum alloy with a liner of cast iron or steel. An alloy having the same rate of expansion as cast iron had not yet been developed, but an aluminum alloy, commercially known as Lo-Ex had been, which had an appreciably lower expansion than the more common alloys. The coefficient of expansion of the Lo-Ex is not the same as cast iron, it is, therefore, desirable to use Ni-Resist iron, which has virtually the same expansion coefficient as the Lo-Ex alloy, for the cylinder sleeves. This material is an austenitic cast iron developed by the International Nickel Company.

Since the piston is both a reciprocating member and a medium of heat transfer, the greatest advantage results in the use of aluminum for this particular part.

In order to use the desired close clearances, the coefficient of expansion of the aluminum piston must more nearly approach that of the material from which the cylinders are made. Using the combination of Lo-Ex and Ni-Resist for the pistons and sleeves eliminates the necessity for piston expansion control, and the use of flexible pistons, because all the materials expand at virtually the same rate. The result is that the piston will be free from the loads caused by excessive piston expansion, eliminating the high friction at elevated temperatures which cause increased wear.

The present engine has been operated for about two hundred hours, using a standard Liberty engine "test club" as load. The engine idles smoothly; has shown no signs of overheating at full load; is exceptionally free from vibration, being perfectly balanced; and accelerates very rapidly and smoothly.

The obvious advantage of the barrel-type engine lies in the greatly reduced frontal area offered by this type as compared with the conventional radial engine of similar power, for example, the Franklin engine rated at 400 h.p. has an overall diameter, including the exhaust ring surrounding the engine, of twenty-three inches. In comparison with this, radial engines of similar power have an overall diameter of fifty-six inches. This, together with the low weight per horsepower and very few working parts, is very interesting to the designer of high-speed planes as they are more or less handicapped by excessive weights and large frontal areas of commercial power plants now on the market.

Other advantages of the Franklin engine are: freedom

from vibration due to perfect balance; extreme light weight per horsepower due to successful utilization of the two-cycle principle which permits more horsepower per cubic inch of cylinder capacity than is the case with four-cycle types. Then, too, we have no valves to consider, resulting in a much better distribution of the cooling medium over the cylinder heads; temperatures are more even within the cylinder; the engine has better cyclic regularity, and the stresses involved are not nearly as severe. The N.A.C.A. conducted tests to determine the efficiency of two-cycle engines, using a single cylinder from a Liberty engine operating on the four-cycle principle they developed twenty-seven horsepower at 1,300 rpm. As a two-cycle, operating on the same principal as developed in the Franklin engine, it consistently developed fifty-three horsepower at the same speed with fuel consumption about equal per BHP hour in both cases. Results of these tests are given in the N.A.C.A. Report No. 239.

Since all accessories are calculated in frictional horsepower, we naturally eliminate a good portion of this by the absence of valves, and valve gear, as it requires considerable horsepower to overcome the resistance of the valve springs and the inertia of the valves. Another source of internal friction which assumes serious proportions is the piston. In the Franklin engine all the thrust is taken in the central portion of the piston, which acts as a connection between the two working heads. This central portion operates between two hardened and ground steel guides having oil delivered to each side at any predetermined pressure, the piston itself, actually never touches the cylinder walls, it merely serves as a guide for the piston rings and has sufficient clearance to take care of any expansion that might occur, as a result of which this point of friction is eliminated.

I hope this brief description, and the photographs, will give the technically minded a better insight into the design and operation of this particular engine.

CHAPTER XIV

Return to Nevada

A FEW DAYS AFTER MY ARRIVAL IN RENO, I WAS FORTU-
nate in meeting a very fine gentleman, F. H. Sibley,
Dean of Engineering at the University of Nevada. We
seemed to enjoy exchanging experiences, and after some
time, as we were having one of our little discussions, he
informed me that he had some bright boys graduating
that year, and, if I could spare the time, he would appreciate it if I could set up a hypothetical project and
take them through it. He felt that, since I had just
come from the East where they were doing things, it
might be of considerable value to them. Sometime before this I had discussed at some length the requirements of aircraft engines with Colonel Page at Wright
Field, and he told me that they were installing dyna-
mometers capable of absorbing five thousand horsepower, even though there was nothing in sight, at that
time, anywhere near that size. He suggested that I
work up a set of drawings for an engine of two thousand horsepower. This, then, gave me a project for
Dean Sibley.

We prepared a complete set of drawings for an
engine of that size, as well as building a small four-cyl-
inder engine of about one hundred horsepower. In this

way the boys got the experience of building a running engine as well as the experience on the drawing board. The ironical part of the whole thing was that I had never gone to college, and yet here I was teaching. I was, of course, a registered engineer, but it had all been learned from books and long hours of study. To say these boys enjoyed the project would be putting it mildly, and many times we found ourselves working well past midnight. But when it was completed and everyone was going home for the summer, the depression had reached the West, and work for engineers was nonexistent, so, like many others, it was necessary for me to change my way of life, as I had a family to take care of and feed.

Fortunately for me, I was able to find employment with the Federal Housing Administration, and served as Field Representative for the state of Nevada and the nine northeast counties of California, which took me as far west as Roseville, and north to the Oregon line. This was a rewarding experience for me, as I covered a very large area, and was to all intents and purposes my own boss, as my superior, after about four months on the job, told me to go where I felt I was needed. I appreciated the confidence he apparently had in me, and tried very hard never to let him down, working long hours and sometimes leaving at 3:00 A.M. in order to be in the next town three hundred miles away when the bank opened at 10:00 A.M. There were, of course, the more pleasant and sometimes humorous sides to my work, and I was able to spend considerable time during the summer months at Lake Tahoe, and more time during the winter months at Las Vegas.

As to the humorous side, I remember one day driving from Reno to Yerington on business, and, having several calls to make, I parked my car and walked to the various places, one of them being the courthouse, where I had some business to transact with one of the officials. After discussing his problem for some time, he insisted that I go out to see his house and continue our discussion there. He was quite badly crippled, and as we descended the steps he approached a new car at the curb, and held the door open for me to enter. Thinking he probably had trouble driving, I said, "Do you want me to drive?" He replied, "Yes," whereupon I moved over under the wheel, and we started out.

Some time later, as we were returning to the courthouse, a red-faced and very angry man pulled alongside and yelled, "Where in hell did you get that car?" I looked at my companion in astonishment and said, "Isn't this your car?" and he replied, "Hell, no, I thought it was yours!" The man alongside happened to be the sheriff, and I had driven off in his car. Fortunately for me it was one of the county officials sitting beside me, as explanations were certainly in order, but we parted as good friends and the drinks were on me.

I enjoyed my work even though it was new to me, and having had some experience with heavy construction, it was not difficult to make the transition to residential work. Actually my work was promotional and educational, but, because of the vast distances to be covered between towns, and the difficulties of sending architectural inspectors to the various jobs, I was quite often called upon to do this work.

During the summer of 1936, I decided to take a few days of my vacation time and visit Captain Eyston, of Britain, who was to be at the Bonneville Salt Flats attempting to set a world speed record. I had a set of drawings we had made at the university with me, and, in going over them, found that there was also a drawing of a power-operated gun turret I had designed some time before. The moment he saw it, he was quite impressed, and said he would like a drawing to take to England with him. The only difference of opinion we had was that he preferred the hydraulic drive, while I preferred the electric. This turret had already been presented to the National Advisory Committee for Aeronautics by Congressman James G. Scrugham with negative results, as shown by a letter mailed to Senator Pittman on July 30, 1935, which follows:

Honorable Key Pittman
United States Senator
Senate Office Building
Washington, D.C.

MY DEAR SENATOR:

Sometime ago, during our conversation in your office at Washington, you mentioned the fact that you would like to have a print of the power operated gun turret for bombing planes which was under discussion at that time.

You will find enclosed print of the turret, together with supporting evidence that it was submitted to the National Advisory Committee for Aeronautics as early as February 26, 1934, by Congressman J. G. Scrugham.

On June 11, 1934, I received a letter from the National Advisory Committee which contained the following: "With reference to armored turrets for airplanes, the increased weight of armor is not condoned by aircraft contractors or the military operating personnel." Having been closely associated with aviation, I, of course, realized that this was true. However, the thought I was trying to convey was not the armor, as they assumed, but the power operated part of the device which made it possible for the operator to do much better work in both aerial photography and gunnery.

September, 1934, I met a Major John G. MacDonnel, U.S.A., Retired. He became very much interested in an aviation engine I have been developing, and requested a set of working drawings for presentation to Lt. Col. H. G. Pratt, Air Corps., Wright Field, Dayton, Ohio.

At that time I handed him a print of the proposed turret to be submitted, and, on November 14, 1934, received an answer from Lt. Col. Pratt containing the following information. "The sketch of the power operated gun turret has been carefully studied, and it is apparently identical with a turret that has been under development by the Air Corps during the last six months. A mockup of such a turret has been constructed, and a machine gun fired from it. The idea appears to be an improvement over existing armament installations, and its application to an actual airplane in the near future is contemplated."

In view of the fact that Lt. Col. Pratt is a member of the National Advisory Committee, it seems hard to reconcile these paradoxical statements. The dates involved are significant and speak for themselves.

I am deeply grateful for the time you so graciously gave me while in Washington, and you may be sure that I shall always be one of your staunch supporters in this area.

Very sincerely yours,

G. E. FRANKLIN

In view of this I told Captain Eyston that, as Great Britain and the United States could not, in my opinion, ever become involved in war, he could take a set of the prints with him, and, after spending a week at Bonneville, observing their preparations to race across the flats which consisted of shimmering white salt, upon which a black line had to be made over the thirteen-mile course as a guide for the driver, I continued on my way to Reno, my home office.

After over three years of service with the Federal Housing Administration, I had traveled the state of Nevada, and the northeast counties of California quite extensively, and had enjoyed the old towns of Placerville (formerly Hangtown), Auburn, Grass Valley, and Nevada City, as well as the northern towns of Westwood, Susanville, and Alturas. Naturally, I enjoyed visiting the old historic towns of the gold-rush days, and even though I worked long hours at my job, it was most enjoyable.

As northern California and Nevada usually had lots of snow during the winter months, traveling was quite difficult, so I spent more time there during the summer when there was greater activity in the building programs. My work also required that I spend quite a lot of time at Lake Tahoe as there was considerable building activity there, too, during the summer. Lake Tahoe was really like a vacation. A beautiful lake at about six thousand feet elevation, completely surrounded with timber, it is truly a thing of beauty, and, as one comes in over the Mount Rose route, it appears like a huge emerald as one tops the crest. The Swiss Alps, with all their claims for beauty, are no more beautiful than Lake Tahoe and the High Sierras.

I did, of course, have to travel the snow country occasionally, and I have bucked snow in second gear for many weary miles.

I also had to go south at times during the summer months, and I experienced the desert heat of Las Vegas.

Even though I had thoroughly enjoyed my work in the mountain, river, and lake areas, I had, through the years, also grown to like the southern part of the state very much. It was entirely different, and yet it had some similarities. The spring, autumn, and winter were ideal, with moderate temperatures, making outdoor living quite comfortable. Then, too, one could broil a steak outdoors, and enjoy eating it, as there were practically no mosquitos or insects of any kind. The summers are quite warm, with temperatures reaching as high as 100° to 110° during July and August, but in thirty minutes one could drive to the mountains, where a heavy sweater is required for comfort. For those interested in fishing, boating, and water skiing, Lake Mead is only thirty miles away, and, for the hunters the surrounding mountains are full of game.

Through the years, I had, because of my work, developed a liking for residential construction, and during October of 1937, I decided to locate in Las Vegas and go into business for myself. This venture was almost immediately successful, and kept me very busy preparing plans and generally supervising the business. In fact, I was so busy that I had almost given up the idea that I might ever be able to revive my plans for continuing the development of the engine, when, during August of 1941, a Mr. Howard Hodson, together with two Air Corps officers whose names I do not remember, set their plane down on McCarran Field, now Nellis Air Base, and came to see me. I had no idea where Mr. Hodson heard of me, or what his connections were, but after some discussion and a demonstration of the engine, it was decided that we should take it to Wright Field. Here, too, luck seemed to be against me, as it was quite obvious that we were heading for trouble with Japan, and those responsible for the defense of the country were already going into production on the major aircraft engines of the United States as well as the Rolls-Royce which was used to power the P-51 Mustang fighter.

By this time I had discovered the hard way that with finances and vested interests being the way they were, a small operator had little, if any, chance of breaking into the manufacture of aircraft engines. However, Mr. Hodson, with whom I became well acquainted, later told me that had we been three months earlier he would have made millions for both of us. He was an unusual man, apparently quite wealthy, as for years he maintained an expensive apartment in New York which was unoccupied most of the time, while he was traveling all over the world.

On the eighteenth day of May, 1945, my first marriage of over thirty years having come to an end some time before, I married my present wife who has, for the past twenty years, been my companion, nurse, and inspiration. Fortunately, our children have, through the years, come to have a deep respect for each other, and they enjoy each other's company when they are together.

Soon after our marriage, we met Hodson's wife, who was a charming person, and his sister, Park. The five of us, Howard, his sister, and our wives, became very good friends. We see them almost every year for a month or so, and never know what to expect when they are

in town. We might open our front door and find Howard standing there with a bottle of champagne in his
hand and saying, "We are coming in and have dinner
with you," or he might say, "Get ready, we are going
out for a show and dinner." At any rate they seemed
to like to be with us, and Persis, his wife, felt quite at
home, playing the piano and singing, which we also
enjoyed. We had letters and cards from them from all
parts of the world, and one day I received a letter asking
us to meet them at the Savoy in London to do England
for a couple of months. Evidently he must have thought
we were about as well off as he was, which, of course,
we were not. My wife and I had a nice visit with them
during October, 1964, at San Carlos, California, where
they had leased a lovely place. We arrived at my broth-
er-in-law's home in San Carlos late in the evening, and
intended to call Howard the following day, but for
some unknown reason I insisted that we call that
evening. It was fortunate that we did, as they were
leaving for Hong Kong at seven the next morning.
Persis answered the phone and Florence said she exclaimed, "Howard, it's the Franklins!" and they insisted that we come right over, which we did, and re-
minisced into the wee small hours. They said they could
catch up on sleep on the boat. We were very sorry to
hear that Park, of whom we were very fond, had died
with a heart attack. She was ninety years of age and
had lived a full and interesting life.

After the episode at Wright Field, I finally realized
that it would be most difficult to raise sufficient finances
to get started again in the aircraft engine field. So I

concentrated on the building industry, which was both enjoyable and profitable, enabling us to have a nice trip occasionally and enjoy our friends.

As my wife and I were both of English ancestry, her mother having come from Birkenhead, a seaport facing Liverpool at the mouth of the River Mersey, where her grandfather had been a marine engineer, we decided we would like to go to England and look up some of our relatives. So, in May of 1953, we started for England to attend the Coronation of Queen Elizabeth, and while there we spent about four months traveling through Europe, a trip which was both educational and enjoyable.

Arriving at Southampton, we disembarked from the *Queen Elizabeth* and boarded the train for London, where we spent two days before going to Edinburgh to tour Scotland. Since all the hotels in Edinburgh were full, we stayed at a beautiful hotel with spacious well-kept grounds about twenty miles from town, called Hydro Peebles, because of the mineral springs there. We visited Edinburgh Castle, where, according to our guide, the crown jewels were on display for the first time in one hundred years. It was at Hydro Peebles that one of the humorous incidents of our visit occurred. We had been making some long trips, trying to see as much of Scotland as possible in a limited time, and, returning just in time to get in the dining room before it closed, all of us wearing our business suits. The Scots, and their ladies, being unable to attend the Coronation in London, were celebrating for several days at Hydro Peebles, and were all in formal attire for dinner and the dance that followed. On Sunday we returned earlier than usual, and decided to wear a tuxedo and evening dress. Imagine our astonishment when we entered the dining room and discovered that we were the only ones in evening clothes. We should have known from our experience on the *Queen Elizabeth* that one does not dress formally on Sunday in that part of the world. However, here we were, so I said to my wife, "This is the time to be nonchalant, my dear." You could almost hear those Scots thinking, "Those crazy Americans!" Whatever they thought of us was quickly forgiven, as we had enjoyed a wonderful week in their country, danced with them in their beautiful ballroom, had a very deep respect for their traditions and love of country, and for their regard for law and order. This was demonstrated every night when, at twelve sharp, the dance band would play their national anthem, and everyone came to attention and then dispersed.

From there we visited the abbeys that played an important part in the border warfare between England and Scotland. Old Melrose was founded during the middle of the seventh century and destroyed in 839. In 1136 King David dedicated New Melrose, which was in turn destroyed about 1545. Holyrood was founded in 1128 and rebuilt in 1220. It is presently the residence of the royal family while in Edinburgh. Dryburgh was founded in 1150, pillaged in 1322, and finally ruined in 1545. It is the final resting place of Douglas Haig, Commander in Chief of the British Forces, World War I. These abbeys were never rebuilt and are now listed among the "Ancient Monuments and Historic Buildings" of Great Britain. The tours of the Scott country, Scott's home at Abbotsford, the lakes, green hills, and heather were truly wonderful.

Our tour of Scotland completed, we returned to London and then went on to Brighton, where I used to go as a boy on vacation. It is a quaint but beautiful town on the south coast of England, about forty-five miles from London, which was as close as we could get during the Coronation, as it seemed everyone was in London.

Very early on the morning of the second of June, we started for London and the Coronation, which was a spectacle one sees only once in a lifetime, if at all. Very colorful it was, with troops from all over the world—white, yellow, brown, and black—and their accouterments were beyond description. The Coronation over, we spent some time visiting relatives, foremost among them being my Aunt Julia who had cared for me when my father died. She had gathered many of my relatives together for an afternoon tea so that they could visit with their cousin from America, and, at eighty-six, she was the life of the party.

After several days visiting in England, we met another group at Harwich, to take a small steamer for The Hague, Netherlands. From there we visited Rotterdam, Amsterdam, Volendam, and across the Zuider Zee to the Isle of Marken, as well as many other places of interest, including the diamond cutters, and flower markets where the flowers were sold at auction, very much like tobacco in the States.

Another interesting thing in Holland was the making

of cheese. Their cow barns were so meticulously taken
care of that one stepped directly into the barn from the
living part of the house, the whole building being under
one roof.

Leaving Holland, we visited Brussels and other points
of interest in Belgium, then on to Aachen, and Cologne
in Germany. Even though the war had been over for
several years, the destruction in Cologne was unbeliev-
able, and our German guide told us the city had been
80 percent destroyed. Their beautiful cathedral was
almost intact, which was understandable, as my son,
who was flying a B-24 bomber, and had flown several
missions over Cologne, told me they had orders not to
bomb it. It was at Cologne that one of our party,
Christine Hubach, who had gone to the United States
when she was quite young and, like us, was returning
to visit relatives in Scotland and England, relieved her-
self of a bitterness she had carried for years. She had
lost two sons in the war and her grief-stricken husband had died shortly thereafter. Consequently, she
was very bitter and seeing bombed-out London did not
help. But in Cologne she asked my wife to go with her
to the cathedral so that she could pray and get the
bitterness out of her heart, because, as she said, "These
people have suffered too." I was more fortunate than
Chris; my son came home even though he was shot
down three times.

Here, at Cologne, we boarded a riverboat and started
up the Rhine, which was the most heavily traveled river
I had ever seen. I had been in the busy ports of the
United States, including New Orleans, but none of them

compared with this for traffic. Arriving at Bonn, we left the boat and spent some time in the city visiting places of interest, including Beethoven's birthplace. Continuing upriver, we came to the Remagen Bridge which had figured so prominently during the war years. Then on to Koblenz, at the confluence of the Rhine and Mosel rivers. We enjoyed this overnight stop very much and visited, among other places, a champagne and wine-making plant, which was most interesting. Some of the world's finest wines and champagnes originate here, being made from Rhine and Mosel Valley grapes.

Continuing upriver, we passed the Lorelei and Mouse Tower before arriving at Rudesheim, where we stayed overnight at an old baronial castle that had been converted into a hotel. Here we left the river and traveled by bus to Mainz, and the old university city of Heidelberg which was most picturesque and interesting. Then we went on to Baden-Baden, a resort town with mineral springs, and through the Black Forest to Freiburg, the last town in Germany. We entered Switzerland at Basel, and continued on to Luzern, Bern, Montreaux, and finally Interlaken, where our group had a very nice birthday party for me, and from where we visited the Swiss Oberland and the Jungfrau, all of them being most beautiful and enjoyable, as were the different quaint little shops in Interlaken itself.

Leaving Interlaken by train, we traveled through the Simplon Tunnel to the beautiful little town of Stresa, on Lake Maggiore, where we stayed a few days, then on to Milan, Venice, Florence, Perugia, Rome, Naples, Sorrento, and the Isle of Capri, spending a few days at each place. Returning to Rome we visited the graves of Keats and Shelley in the Protestant cemetery, then continued up the west coast to Pisa, Genoa, Rapallo, and Portofino, spending a few days at these places to see other points of interest.

Before entering France we stopped at Monaco to see the games at Monte Carlo, then on to Nice, which we enjoyed very much as the Riviera is quite beautiful, and we liked walking along the Promenade des Anglais, built about 1822 at the expense of the English Colony and having two thirty-three-foot-wide carriage ways separated by banked flowers. From Nice we traveled along the coast to Toulon and Marseille, where we turned north to Lyon, Dijon, and Paris. Here we spent ten wonderful days before boarding the boat train for Cherbourg. After traveling in a number of countries and meeting members of several nationalities, I concluded that the reason many, not all, Americans do not fully enjoy a trip to Europe is that they expect to find all the conveniences they have at home, which they do not, so they adopt a superior attitude which Europeans resent. We were most fortunate in having a wonderful guide in the person of Ruth Zurbukken. Ruth was married to a Swiss engineer who had later become an American citizen. She spoke about five languages and seemed to enjoy our company as well as that of a young lady, Bernice Futterer, from Sacramento, California, who was a member of our party. Bernice was an accomplished musician and occasionally entertained for us at the different hotels along the way. So, since we got along so well, we would, quite often, after the rest of our group had retired, go out on the town and mingle with the local people, where we were well received and made to feel at home. In Paris, after our group had broken up and we were on our own for ten days, we learned to use the Metro System and traveled everywhere just like a couple of Parisians. The food was good and we found a little family restaurant, owned and operated by a French family, that served wonderful ham and eggs, which we had been told were unobtainable in Europe, and which we were starved for, having had none since leaving home.

Another place we enjoyed was a restaurant on the Eiffel Tower just above the arch and about one hundred and eighty feet above the pavement. The food was excellent, as was the view of Paris, which we enjoyed while eating our dinner. We also enjoyed the sidewalk cafes on the Champs-Élysées, the Louvre, the flying buttresses of Notre Dame, the book boxes along the Seine, Montmartre (with the beautiful white Basilica of Sacre-Coeur in the background), the Opera House, the Tuileries Gardens, and, of course, the Flea Market at Clignancourt, as well as many other places too numerous to mention. Our ten-day Paris vacation passed all too quickly and we had to take the boat train for Cherbourg to board the *Queen Mary* for our return home.

It was while making preparations for this trip that I discovered I was not considered a citizen by the Bureau of Immigration and Naturalization. In fact they informed me that, as far as they were concerned, I was just a "damned alien." Actually I was a citizen as I had entered the United States before the immigration and naturalization laws were passed, but I could not get an affidavit to support this. Then, too, my stepfather had obtained his final papers before I reached the age of twenty-one, and I had been advised during my twentieth year that I was a citizen through my parents' naturalization. So through the years I had signed that way (citizen by parents' naturalization). I had been checked by the Federal Bureau of Investigation on several occasions, and had served on many critical projects, but getting a passport was another matter, when I discovered that my stepfather, in making out his citizenship papers, had mentioned only his own children, so I was left out. Through the years I had been voting, paying my income taxes, and generally behaving as a citizen, which I thought I was. So it was quite a shock to find that I had to go through the process of becoming a citizen after all these years.

Upon my arrival at the office of the Bureau of Immigration and Naturalization I found that I was being subjected to the usual bureaucratic red tape, and getting nowhere. My tickets for the trip were already purchased and time was running out, so my wife called Miss Eva Adams, Executive Assistant to Senator Pat McCarran, and an old friend, telling her about our problems, and they were straightened out in short order. Going through all this did clear up one thing about which I was beginning to have some doubts, and that was my age when I came to the United States. It did not seem possible that I could have been so young. However, when the examiner questioned me on this point, and I told him that, if my memory was correct, I arrived on January 6, 1903, he pulled some papers from his desk, and said, "That is correct. You disembarked from the S.S. *Cymric*, at Boston, January 6, 1903."

In spite of the seriousness of this problem, it did have its humorous moments. During the examination I was taken in for questioning, and the first question was, "Who is the governor of the state of Nevada?" and, even though I knew the governor quite well, I had a mental block and could not say his name. Finally, the examiner said, "Does Russel mean anything to you?" This broke the spell, and I replied, "Of course, Charles Russel." From then on I had no problems. Then my two witnesses were called in. They were old friends and had taken time off from their businesses to do this for me. Harry Manente, manager of the Third Street Branch of the First National Bank, and Ray Thorn, who had been quite ill, and got out of bed to be there, was manager of the Sears, Roebuck store, or, as he called it, "the big store." These two came out from their interview with long faces, telling me how sorry they were. Finally I asked them what was wrong, and they said, "Well, everything was all right until this communism thing came up, then what could we do."

Their little joke over, they left the building, and I prepared to be sworn in. A fellow Kiwanian, Judge Henderson, with whom I had been swearing allegiance to the flag at our weekly luncheons for the past ten or twelve years, officiated and, I am sure, was glad to make it official.

Upon our return from Europe, I settled down again

into our regular business routine, and, upon retirement recently, had built well over fourteen hundred homes, as well as many commercial buildings. About the time of my retirement, I received a call from an old friend in Montana, Frank W. Wiley, advisor to the Montana Aeronautics Commission. I had not heard from Frank during the past thirty years, but he was one of the old-timers who used to fly by the seat of his pants, into and out of more canyons in the West than any other pilot I knew. He had been appointed to set up an Avia-tion History Project for the state of Montana, and had contacted me to get the history of my engines, and, if possible, the engines themselves. After some discussion and several letters, he suggested that since they orig-inated in Montana, and, even though I had built one of them at the University of Nevada, we had no Aero-nautics Commission, so they should be in Montana, to which I agreed. A few days later a commission truck arrived in Las Vegas to transport them to Helena, Mon-tana, where they are now a permanent part of the aviation history of the state.

Now that I am retired it is comforting to be able to live among my many friends who have contributed so much to this area, among them, Archie C. Grant, under whose leadership as chairman of the Board of Regents, of the University of Nevada, starting from scratch in 1957, has advanced Nevada Southern University until it is now a fully accredited degree-granting university with a campus comprising two hundred acres and seven buildings valued at approximately seven million dol-lars. A U.S. Radiological Laboratory on campus is a definite asset, as is the fine group of top engineers and scientists cooperating with N.S.U. and giving of their time and experience.

The top musicians, both vocal and instrumental, have been so generous with their talents in giving free concerts almost every Sunday under the able direction of Sonia Cobb, Jim Clark, Antonio Morelli, and others. And I would be remiss indeed if at this time I did not think of my dear friend of many years, George Probasco. I have valued his friendship, counsel, and advice for many years and I am glad to know that he and his good wife Peggy are presently enjoying themselves traveling in Europe.

As for me, it has been my pleasure and privilege through the years to be president of the Builders Exchange, Associated General Contractors, Real Estate Board, Las Vegas Knife and Fork Club, and I am presently a licensed mechanical engineer, a member of the National Society of Professional Engineers, and the treasurer of my Masonic Lodge, a post I have held for several years.

In conclusion, if accumulating a lot of worldly goods means wealth, then I am not a wealthy man, but if living a full and interesting life, taking a little of the bitter along the way in order to appreciate the sweet, living in a modest but lovely home, supervised by a charming, talented, and capable wife, and being surrounded by wonderful friends means wealth, then I am rich indeed, and, as I approach the sunset years, I am content.